T0177495

Agency in Mental Disorder

Agency in Mental Disorder

Philosophical Dimensions

Edited by
MATT KING & JOSHUA MAY

Great Clarendon Street, Oxford, OX2 6DP,
United Kingdom

Oxford University Press is a department of the University of Oxford.
It furthers the University's objective of excellence in research, scholarship,
and education by publishing worldwide. Oxford is a registered trade mark of
Oxford University Press in the UK and in certain other countries

© the several contributors 2022

The moral rights of the authors have been asserted

First Edition published in 2022

Published in the United States of America by Oxford University Press
198 Madison Avenue, New York, NY 10016, United States of America

British Library Cataloguing in Publication Data
Data available

Library of Congress Control Number: 2021940784

ISBN 978-0-19-886881-1

DOI: 10.1093/oso/9780198868811.001.0001

Printed and bound in the UK by
TJ Books Limited

Contents

Preface

According to the National Institute of Mental Health, roughly one in five adults in the U.S. live with a mental disorder. Its prevalence, not just in America but around the world, makes it an issue of great practical importance, studied by psychologists, neuroscientists, and psychiatrists. Philosophers, however, have given it less sustained inquiry. While there is some work that touches on the significance of mental illness to agency and responsibility, and a few extended treatments of particular disorders, it only scratches the surface. Most often, mental disorders are used as illustrative examples, as a means to some other argumentative end, rather than to better understand the disorders themselves and their effects on human agency. Our volume seeks to broaden and deepen the philosophical analyses.

Our goal was to produce a book that will be attractive to more than just philosophers. The relationship between agency and mental illness is an area of inquiry with numerous practical implications and is of interest to academics in cognate fields, such as law, neuroscience, and clinical psychology, as well as those that practice in such fields (e.g., psychiatrists, lawyers), and to the general public at large. The contributions in this volume, despite being all from philosophers, are nonetheless aimed at this broader audience.

The initial idea for the theme of the volume arose out of a project on responsibility and mental disorder, in which we co-authored an article defending the need for a more nuanced understanding of that relationship. That paper, which was largely programmatic, seemed to resonate with many working on the topic. And we noticed in our research that there was a dearth of dedicated treatments with both the sort of breadth and attention to empirical details that we thought necessary.

So, in March 2018, we gathered a number of philosophers working in empirically informed ways on agency and responsibility for a two-day workshop at the University of Alabama at Birmingham. The result was a number of very interesting and provocative presentations, pushing new ideas and questions, most of which have now found their way into the pages that follow. The participants were Nomy Arpaly, Justin Clarke-Doane, Anneli Jefferson, Lauren Olin, David Shoemaker, Walter Sinnott-Armstrong, Chandra Sripada, Jesse Summers, Kathryn Tabb, and Natalia

Washington, and were joined by some wonderful local philosophers, including, Marshall Abrams, Holly Kantin, Rekha Nath, Jason Shepard, Mike Sloane, and Mary Whall. Our thanks to both groups. Their participation made for a lively exchange of ideas in the true spirit of a workshop, wherein ideas are tested and revised in a friendly and cooperative atmosphere.

Thanks to the Provost's Office, Dean's Office, and Department of Philosophy at UAB for a Faculty Development Grant supporting the workshop. May's work on this volume was also made possible through the generous support of a grant from the John Templeton Foundation (Academic Cross-Training Fellowship 61581). The opinions expressed in this publication are those of the authors and do not necessarily reflect the views of the John Templeton Foundation. Thanks as well to three industrious philosophy majors—JaVarus Humphries, Campbell Mackenzie, and Mohammad Waqas—as well as to our departmental administrative associate, Donna Miller, for providing logistical and administrative support. Finally, thanks to Peter Momtchiloff at Oxford University Press for his support of the project and editorial advice.

Notes on Contributors

Nomy Arpaly is Professor of Philosophy at Brown University.

Justin Clarke-Doane is Associate Professor of Philosophy at Columbia University.

Anneli Jefferson is Lecturer in Philosophy at Cardiff University.

David Shoemaker is Professor of Philosophy at Cornell University.

Katrina L. Sifferd is Professor of Philosophy at Elmhurst University where she holds the Genevieve Staudt Endowed Chair.

Walter Sinnott-Armstrong is the Chauncey Stillman Professor of Practical Ethics in the Department of Philosophy and the Kenan Institute for Ethics at Duke University.

Chandra Sripada is Associate Professor of Psychiatry and Philosophy at the University of Michigan.

Jesse S. Summers is Assistant Academic Dean, Trinity College of Arts and Sciences; a Fellow at the Kenan Institute for Ethics; and an Adjunct Assistant Professor of Philosophy at Duke University.

Kathryn Tabb is Assistant Professor of Philosophy at Bard College.

Robyn Repko Waller is Assistant Professor of Philosophy at Iona College.

Introduction

Matt King and Joshua May

Our understanding of mental disorders is undergoing a paradigm shift. Historically, many patients thought to be afflicted with mental illness were placed at the margins of society, if not entirely excluded, based on the assumption that they are incompetent or lack full autonomy. A richer understanding of mental disorder, however, has generated alternative perspectives. Some theorists have argued that only certain patients have limited forms of autonomy (e.g., Pickard 2015; Kozuch & McKenna 2015; Shoemaker 2015). Other views are more radical, suggesting that some patients' agency is not necessarily disabled but merely different; atypical, if not neurotypical (see e.g. Glannon 2007; Silberman 2015; King & May 2018).

Such perspectives raise critical philosophical questions about free will and responsibility. How exactly do mental disorders affect one's agency? How might therapeutic interventions help patients regain or improve their autonomy? Do only some disorders excuse morally inappropriate behavior, such as theft or child neglect? Or is there nothing about having a disorder, as such, that affects whether we ought to praise or blame someone for their moral success or failure?

Our volume gathers together empirically informed philosophers who are well equipped to tackle such questions. Contributors specialize in free will, agency, and responsibility, but they are informed by current scientific and clinical approaches to a wide range of psychopathologies, including autism, addiction, personality disorders, depression, dementia, phobias, schizophrenia, and obsessive-compulsive disorder. These conditions exhibit a diverse array of symptoms that can contribute quite differently to blameworthy or praiseworthy acts.

Prior philosophical work has tended to only briefly address the significance of mental illness to agency and responsibility. One can find some extended treatments of particular disorders, such as the appropriateness of

Matt King and Joshua May, *Introduction* In: *Agency in Mental Disorder: Philosophical Dimensions.* Edited by: Matt King & Joshua May, Oxford University Press. © Matt King and Joshua May 2022.
DOI: 10.1093/oso/9780198868811.003.0001

blame for individuals with psychopathy or those grappling with addiction. But these treatments exclude a wide variety of other rather different psychopathologies. Our collection of essays seeks to expand the conversation and push theorizing in new directions.

In this opinionated introduction, we not only summarize but contextualize the chapters. Our aim is to form a cohesive collection of philosophical essays that serves as a foundation for ongoing connections between theory and clinical practice. To that end, we identify two key approaches to our topic (theoretical vs. therapeutic), as well as some common framing assumptions that end up being scrutinized or defended by our contributors. Finally, we briefly conclude that rejecting or revising the framing assumptions may help thread the needle between empowering patients who suffer from mental illness while avoiding the stigmas that add insult to injury.

0.1 Theoretical vs. Therapeutic Perspectives

We can begin with a rough distinction between two perspectives on disorder. The *theoretical perspective* sees a disordered system as, in some sense, malfunctioning. Emphasis here is on what the system is, how it works, what its true nature is. Such an approach might treat mental disorder on an analogue with other forms of malfunction. The typical idea here would be to understand what well-functioning consists in, treating psychopathology as importantly deviating from that standard. Something has gone wrong; perhaps something needs to be fixed.

Contrast this with the *therapeutic perspective*. This perspective sees a person in distress, and so disorder is understood in terms of its adverse effects on the individual. Here, if we appeal to a standard, it is an evaluative one, concerned with how well the person's life is going *for them*. As such, the goal is to improve the patient's life by utilizing techniques to manage or mitigate those adverse effects through the patient's own agency.

These two approaches to disease and disorder are of a sort familiar from philosophical discussions. Compare, for example, biomedical versus social constructionist accounts of mental disorder (Graham 2010) and naturalistic versus normative accounts of disease generally (Stegenga 2018). Such approaches needn't be mutually exclusive, but in philosophical discussions one approach is sometimes emphasized to the neglect of the other. Though a rough approximation, the theoretical perspective has dominated philosophical discussions of mental disorder (at least in the analytic tradition), while the therapeutic perspective dominates most clinical settings, particularly those of social workers and psychotherapists. We offer these contrasting perspectives as a useful device for better appreciating the broader context in which we might evaluate the relationship between agency and mental disorder. Several of our chapters bring them together, and we conclude this introduction with an attempt to articulate how theoretical issues can inform therapeutic concerns.

0.2 Framing Assumptions

A hallmark of philosophical analysis is the unearthing and evaluation of latent assumptions. Unsurprisingly, there are a number of assumptions about agency and mental disorder that bear careful consideration and scrutiny. We do not insist that the following assumptions have been held explicitly or by all theorists, but they appear in some common strands of theorizing, clinical approaches, and public discourse. As we'll see, each framing assumption helps to justify attributing less agency to those with mental disorders, which affects both how we understand the nature of mental disorder and how we respond to those with particular disorders.

Categorical assumption—Mental disorders exempt or mitigate categorically, similar to how being a child should exempt one from military conscription. In virtue of having a mental disorder, one's responsibility is thereby mitigated in some way. Even if disorders exist on a spectrum with neurotypical individuals, the disorders deviate greatly from the norm, and thus they differ significantly from normal populations. Just as maturity comes in degrees, yet we still treat childhood as categorically distinct from adulthood, at least in certain contexts, such as legal culpability.

Reduction assumption—Mental disorders are better understood as brain disorders. We come to better understand the nature of a mental disorder if we can uncover the neurological mechanisms involved. Patients in the grips of addiction, for instance, are said to have a brain disease, which helps to ground a disease model according to which patients lack control over their urges to use or to procure the means to using. We also see this assumption arise in the context of research and its funding. In the United States, the National Institute of Mental Health now encourages researchers to connect their studies and hypotheses with biological concepts from fields like neuroscience and genetics (based on the Research Domain Criteria).

Passivity assumption—Mental disorders cause things that happen to us rather than involve things that we *do*. The characteristic features of mental disorders include things like urges, spasms, delusions, etc., processes that seemingly occur *to* a person rather than being products *of* a person. Symptoms are thus easily cast as compulsions or irresistible thoughts and desires that arise unbidden.

Internalist assumption—Mental disorders are fundamentally "inside" a person. Rather than driven by outside forces or individuals, features of mental disorders are internal to the patient's psychology. In the past, the assumption has been roundly stigmatizing: "It's all in your head." However, more compassionate approaches still often aim to improve patients' symptoms by focusing only on changing their thoughts or desires, through e.g. counseling, pharmaceutical drugs, or brain stimulation.

Most of our contributors ultimately put pressure on one or more of these assumptions, though the Categorical assumption receives the least support. However, at least one of our contributors (Chandra Sripada) can be read as defending many of these assumptions, drawing on clinical and experimental evidence. What emerges in this volume, then, is a lively debate about which theoretical frameworks we ought to use when addressing questions of agency in psychopathology.

0.3 Chapter Summaries

The somewhat disparate nature of the topic at hand resists any straightforward organization of the contributions. Instead, we've ordered the chapters along two main dimensions. First, virtually all of the chapters critically engage with one or more of our framing assumptions, though they all do so implicitly. The exception is the final chapter, which we see as providing a contrasting defense of the framing assumptions. (As a reminder, the framing assumptions are *our* way of providing a context and thematic organization to the guiding questions of the volume, rather than being the explicit targets of our contributors.) Second, and to a somewhat rough approximation, we've otherwise ordered the chapters with a view toward how general their topic is. Granting an interest in the relation between agency and mental disorder, one's focus might nevertheless concern elements at varying levels of specificity. Our opening chapters engage with that relationship in more general ways, while later chapters focus on more narrow questions—though this doesn't prevent them from drawing quite general lessons.

Apart from these very rough principles, there is no further way in which the entries can be easily grouped. (Perhaps it would be more accurate to say there are *many* alternative ways in which one could group the chapters.) But this is a virtue of the volume. Our aim is to initiate and stimulate new directions of inquiry regarding the relationship between agency and mental disorder. Our contributors have written chapters that reflect the rich and complicated nature of the topic. We summarize each below, relating its central ideas to our framing assumptions.

In Chapter 1, Nomy Arpaly examines how a "quality of will" view of moral responsibility might apply to the sorts of "unusual behaviors" that can arise in the context of mental disorder. One's blameworthiness and praise-worthiness, on her view, are determined by whether one's behavior reflects a proper concern for moral reasons. A person's act is blameworthy, for example, if it reflects inappropriate malice towards others or indifference to their well-being. Although psychopathologies might reliably yield unusual behavior, they do not always compromise or even impair one's ability to grasp and respond to moral reasons, either appropriately or inappropriately. Although psychopathologies can make it difficult to understand a patient's true motivations or reasons for acting, this does not necessarily imply an inability to grasp or respond to reasons, even if we classify the affliction as a disease or disorder. Pathologies, on her view, can reduce one's responsibility, but only due to specific symptoms that stand in the way of good will. Arpaly's discussion traverses a wide range of issues, drawing numerous profitable distinctions along the way regarding delusions, irrationality, depression's variable effects, and narcissism.

In perhaps the strongest rejection of the Categorical assumption, Arpaly dismisses the concept of mental disorder (as it is typically employed) as "not philosophically respectable." She notes that the boundaries of particular disorders have more to do with practical concerns than theoretical ones. What matters for responsibility is how and in what ways one's concern for one's moral reasons might be affected, and this relationship is unlikely to follow the diagnostic categories in any direct way.

In Chapter 2, David Shoemaker tackles an underexplored moral *risk* associated with denying responsibility for those with mental disorders. Many theorists have followed P.F. Strawson's influential account linking our practices of holding each other accountable with treating each other as persons toward whom it makes sense to feel and express reactive attitudes such as resentment and admiration. But, if this thought were correct, those exempted from accountability would therefore be excluded from the community of persons and thus certain social goods. Shoemaker argues that this exclusion on the basis of arbitrary characteristics is discriminatory, labeling such exclusions as *accountability-disabilities*. Arguing that the two prevailing models of ameliorating these burdens of disability fail, Shoemaker instead negotiates the problem by denying the common theoretical assumption that holding one another accountable requires treating each other as persons. Separating the two domains of accountability and interpersonality allows for exempting some patients from responsibility without thereby excluding them from other important and valuable aspects of our interpersonal relations. Dividing the two domains also points to interesting insights about our shared humanity.

At its heart, Shoemaker's chapter critically examines the Categorical and Internalist assumptions, asking us to more fully consider their implications. How central are our responsibility practices to our shared lives as humans? The resulting discussion explores the way in which our theories of mental disorders and responsibility can make life worse for those with them, but also how theory can help.

In Chapter 3, Anneli Jefferson examines how brain abnormalities, specifically, affect a patient's moral responsibility. Immoral acts can seem less blameworthy if we discover that the perpetrator suffered from a brain tumor or neurodegenerative disease. Jefferson argues that such neurobiological facts are at best indirectly relevant to one's responsibility. When reduced blame is appropriate in such cases, it is not simply because the patient had a brain abnormality but rather because that abnormality impaired psychological capacities, such as comprehension of what one is doing or control over one's actions. Jefferson argues that brain disorders might seem to mitigate blame only because they often cause radical changes in one's character traits, which impair capacities like self-control by nullifying one's developed strategies and previously established support networks for managing wayward thoughts and desires.

Jefferson's chapter can be seen as questioning both the Reduction and Internalist assumptions, given her insistence that, when it comes to agency and responsibility, the psychological level of explanation remains primary, which is intertwined with one's social conditions.

In Chapter 4, Robyn Repko Waller brings key psychiatric concepts into direct conversation with a popular philosophical approach to moral responsibility, to better understand how mental disorders can affect agency. An important element of our responsible agency involves control: we are only responsible for our actions to the extent that we can exercise appropriate control over them. One prominent philosophical model for such control understands it in terms of an individual's reasons-responsiveness. We exercise control through our ability to recognize patterns of reasons for action and align our conduct with those reasons. Waller then examines effective psychotherapies and argues that they work by enhancing patient control by improving their reason-responsiveness. In particular, she looks at Exposure Therapy as a treatment for agoraphobia, and Dialectical Behavior Therapy for patients with borderline personality disorder. By applying the model of reasons-responsiveness, Waller argues we can better understand key psychiatric concepts, like distress tolerance and experiential avoidance, and the role they play in agential control more generally.

Waller's chapter most directly challenges the Passivity assumption, arguing that standard psychiatric therapies work precisely by targeting the patient's rational capacities. Additionally, to the extent that external elements, such as one's environment, can impact reasons-responsiveness, her discussion casts some doubt on the Internalist assumption as well.

In Chapter 5, Katrina Sifferd explores the role of mental disorders in criminal liability. The criminal law recognizes a special excuse for legal insanity: if defendants can't appreciate the wrongfulness or criminality of their conduct, then they are not guilty even if they otherwise satisfy the criteria for criminal liability. Sifferd tries to make sense of this special excuse in light of the fact that, ordinarily, having false moral or legal beliefs isn't a recognized criminal defense. Sifferd argues that there is no easy move to make from a diagnosis of mental disorder to reduced criminal liability. Instead, it is necessary to evaluate the ways in which the symptoms of the disorder affected the defendant's action at the time. The role of the legal insanity excuse is to spare those who lack an understanding of the moral or legal nature of their situation and who are not culpable for this lack. Thus, symptoms of a mental disorder are relevant because they can sometimes cause a lack of moral knowledge, and a diagnosis of a mental disorder *signals* to the court that such ignorance might be present.

Despite being focused on the criminal law, Sifferd's chapter provides further insight on the Categorical and Internalist assumptions. The legal insanity excuse does not provide a global exemption, on her view, but rather must be established in the context in which the crime was committed. Moreover, to assess a defendant's moral knowledge or ignorance, we must take into account the ways in which their environment interacts with their cognitive capacities.

In Chapter 6, Jesse Summers and Walter Sinnott-Armstrong discuss a version of obsessive-compulsive disorder (OCD) known as Scrupulosity. Patients with this disorder exhibit obsessions with avoiding unethical behavior or doing what is morally or religiously sanctioned, at least as they interpret the relevant rules. These obsessions lead to compulsions, such as repeatedly checking that one has not inadvertently poisoned a co-worker, which reduce the patient's anxiety. As with other forms of OCD, the obsessions and compulsions can yield an intense focus that leads patients to neglect other moral duties, such as parental care. Are these patients fully responsible for moral failings caused primarily by their compulsions? Summers and Sinnott-Armstrong argue that patients with Scrupulosity are usually less responsible, because their extreme anxiety makes it difficult to be responsive to reasons to act otherwise, similar to how a back injury forces a parent to blamelessly miss a child's piano recital. The authors draw analogies between Scrupulosity and some cases of addiction in order to reinforce classical objections to a main rival—so-called Deep Self Views—which might otherwise seem to serve as a better explanation of responsibility in cases of Scrupulosity.

Ultimately, Summers and Sinnott-Armstrong's chapter might raise doubts about the Categorical and Passivity assumptions. Although they regard many patients with Scrupulosity as blameless, they analyze the responsibility of these patients in terms of degrees. Moreover, the authors describe obsessions in Scrupulosity as being active attempts to reduce one's anxiety, not a passive urge that arises spontaneously.

In Chapter 7, Justin Clarke-Doane and Kathryn Tabb critique the use of the exemplar of the addict to inform theories of free will, praise, and blame. The case of the addict is often used in philosophical arguments to help demarcate which psychological influences on one's action mitigate responsibility for it. The characteristics of addiction are an empirical matter, however, and often there is an assumption that science can help reveal what it is about the cravings and compulsions of addiction that compromise agency, unlike ordinary behavior. Yet scientific accounts of addiction seem to uncover neurobiological mechanisms that are different only in degree, not in kind, from non-addicts who engage in blameworthy behavior, such as theft, infidelity, or even procrastination. Clarke-Doane and Tabb conclude, first, that we should be doubtful that theorizing about responsibility will be advanced by focusing on particular kinds of psychopathologies, such as addiction but also Tourette syndrome, OCD, and the like. Second, the authors suggest that perhaps the way forward is to accept the psychological similarities between addictive and non-addictive behavior and simply treat like cases alike. Either treat addicts as we normally treat non-addicts (as fully culpable) or embrace the skeptical conclusion that everyone is less responsible than we thought—perhaps not responsible at all.

Clarke-Doane and Tabb can be seen as questioning our Reduction and Categorical assumptions. Their chapter cautions that the science, including neurobiological explanations, of addiction (and possibly other psychopathologies) do not necessarily help to uncover differences in human agency that seem intuitively present. In particular, neurobiological explanations do not justify the portrayal of addiction as a brain disease, at least not one that causes the actions of addicts to be categorically different from non-addicts.

Finally, in Chapter 8, Chandra Sripada contends that mental disorders do substantially diminish patients' control over their actions, compared to neurotypical individuals. The key, he argues, is to reject a standard model of human action which says that one's behavior is always driven by one's strongest desires. Instead, humans have default habits and impulses that generally produce corresponding behavior, unless they are inhibited or regulated. Yet regulatory control often fails, particularly in the long-term. Over hours or weeks or months, patients chronically experience deviant impulses that become increasingly difficult to inhibit, and this makes brute failures of control more likely. Although neurotypical individuals similarly struggle to pay attention or stay on task as distractions pile up throughout the day, Sripada argues that those with psychopathologies—like depression, OCD, ADHD, and schizophrenia—experience many more impulses over time that become impossible to consistently regulate. Sripada thus forcefully rejects "volitional" views according to which patients with mental disorders freely choose to engage in aberrant behavior because they ultimately want most to do so (despite repeatedly seeking clinical help).

Sripada's chapter can be seen as defending all four of our framing assumptions, in light of evidence from cognitive and clinical neuroscience. First, although the regulatory control framework puts those with psychopathologies on a continuum with the neurotypical, Sripada clearly views mental disorders as generally yielding *categorically* greater limits on control. (Recall our analogy: although the maturity of teenagers differs only in degree from that of adults, we still treat one as categorically distinct from the other.) Second, Sripada finds it fruitful to *reduce* the regulation of complex actions to simple forms of cognitive control studied in neuroscience. Finally, Sripada locates the crux of diminished agency in the patient's *internal* states, such as impulses or urges, which are also treated as largely *passive* and unrepresentative of the patient's true values. In this way, Sripada's chapter serves to revitalize and fortify the framing assumptions that some other chapters attack.

0.4 From Theory to Practice: Stigma

In the early days of psychiatry, patients were often treated unethically. They were ridiculed, ignored, experimented on without consent, and cast to the margins of society. Mental healthcare has come a long way in its efforts to be more compassionate toward patients. Many of us—friends, family members, and patients themselves—have likewise sought to be more empathic and understanding. The chapters in this volume are largely theoretical discussions, but they have practical, even therapeutic, implications. We close this introduction with one important potential impact of theory on patients suffering from mental illness: stigmatization.

Stigma toward mental illness comes in many forms, including the stereotypes that such patients are dangerous, incompetent, or weak in character (Rüsch et al. 2005). Many of these stereotypes link directly to responsibility. As several of our contributors emphasize, competence and self-control are core forms of agency necessary for autonomy and for the fittingness of praise and blame.

Stigma matters for several reasons, but we'll focus on two. First, fear of stigma can decrease a person's motivation to seek mental health services, to pursue employment, or to live independently (Rüsch et al. 2005). Second, as Shoemaker's chapter emphasizes, removing agency from another person, or significantly diminishing it, can exclude them from basic social interactions that humans value and need for the maintenance of mental health. In other words, stereotypes of reduced agency can lead to prejudice, and even discrimination, whether it arises from others or oneself. Thus, what might seem like merely academic concerns about agency can have deleterious effects on the lives of patients and those who love and care for them.

How might theoretical considerations reduce, rather than exacerbate stigma? One route to destigmatization generally involves diminishing a person's role in the negative consequences of one's action. For example, we might seek to destigmatize poverty by noting the various external causes and structural forces that lead to economic hardship for families. Their condition isn't their fault, so to speak, and so attaching negative attitudes to such a condition is to treat them unfairly.

The same strategy can be applied to mental disorder. It's not her fault; she can't help it; it's a compulsion; he has a brain disease. The framing assumptions help to justify such talk. If psychopathology is construed as something passive that occurs entirely inside the patient that can be reduced categorically to a malfunction of the brain, then this might reduce stigma by making blame inappropriate. But destigmatizing in this way may come at the cost of disempowering patients and keeping them on the margins as categorically different from neurotypical individuals. Indeed, although societies have increasingly embraced a more neurobiological conception of mental disorder, some evidence suggests that this has led to increased stigma (Schomerus et al. 2012).

Scrutinizing the framing assumptions is thus critical for understanding how to approach agency in mental disorder. The assumptions generally accord patients less agency, which helps support a compassionate, therapeutic perspective on their pathology—patients deserve treatment, not blame. Yet reduced agency is a mixed blessing: it can mitigate blame at the cost of further marginalizing patients grappling with mental illness (Haslam & Kvaale 2015).

One might worry that ascribing too much agency in mental disorders will re-stigmatize and reinforce marginalization. However, this seems inevitable only if we retain the framing assumptions that are scrutinized (though sometimes defended) in this volume. Rejecting or revising the framing assumptions might provide theoretical room for both empowering patients as agents and reducing stigma by emphasizing similarities with the neurotypical (Arpaly 2005; Glannon 2007; Kennett 2007; Pickard 2015; Kozuch & McKenna 2015). Some theoretical support thus arises for an approach to psychopathology that is able to attribute more agency while reducing stigma. Although the two aims may initially seem in conflict, there may be philosophical grounds for reconciling them.

Indeed, modifying the framing assumptions might be integral to a compassionate, therapeutic approach to psychopathology. Often treatment of mental disorder requires taking responsibility for one's actions. Clinicians cannot, in the end, force patients to behave differently. The more agency that is stripped from patients, the less improvement one can expect. Nevertheless, as Hanna Pickard (2011) has put it, one can accord "responsibility without blame." Imputing agency doesn't require chastising, for example, which can alienate patients from their treatment plan.

None of this is to deny that psychopathologies can cause mental anguish and sometimes insurmountable obstacles to the full exercise of agency. Yet people with autism, dementia, major depression, even schizophrenia can lead lives filled with responsibilities toward family members, friends, and their community that require the exercise of human agency. Often these success stories are dependent upon an active treatment plan, but even then the mental disorder is rarely cured but instead actively managed. In some cases, mental illness involves or develops such severe symptoms that one's agency is decimated or entirely lost, as in the late stages of Alzheimer's. But such cases do not necessarily represent the norm, and their existence does not justify generalizing to all psychopathologies. Perhaps little can be said in general about agency in mental disorders that will apply to all or most of them (King & May 2018). Realizing the ways in which our theorizing must be more nuanced is precisely the sort of philosophical progress we hope this volume spurs.

References

Arpaly, N. (2005). How it is not "just like diabetes": Mental disorders and the moral psychologist. *Philosophical Issues* 15(1): 282–98.

Glannon, W. (2007). Neurodiversity. *Journal of Ethics in Mental Health* 2(2): 1–5.

Graham, G. (2010). *The Disordered Mind: An Introduction to Philosophy of Mind and Mental Illness*. New York: Routledge.

Haslam, N., and Kvaale, E. P. (2015). Biogenetic explanations of mental disorder: The mixed-blessings model. *Current Directions in Psychological Science* 24(5): 399–404.

Kennett, J. (2007). Mental disorder, moral agency, and the self. *The Oxford Handbook of Bioethics*, ed. Bonnie Steinbock, 90–113. Oxford: Oxford University Press.

King, M. and May, J. (2018). Moral responsibility and mental illness: A call for nuance. *Neuroethics* 11(1): 11–22.

Kozuch, B., and McKenna, M. (2015). Free will, moral responsibility, and mental illness. In *Philosophy and Psychiatry: Problems, Intersections, and New Perspectives*, eds. D. D. Moseleyand G. Gala. New York: Routledge.

Pickard, H. (2011). Responsibility without blame: Empathy and the effective treatment of personality disorder. *Philosophy, Psychiatry, & Psychology* 18(3): 209–23.

Pickard, H. (2015). Psychopathology and the ability to do otherwise. *Philosophy and Phenomenological Research* 90(1): 135–63.

Rüsch, N., Angermeyer, M. C., and Corrigan, P. W. (2005). Mental illness stigma: Concepts, consequences, and initiatives to reduce stigma. *European Psychiatry* 20(8): 529–39.

Schomerus, G., Schwahn, C., Holzinger, A., Corrigan, P. W., Grabe, H. J., Carta, M. G., and Angermeyer, M. C. (2012). Evolution of public attitudes about mental illness: A systematic review and meta-analysis. *Acta Psychiatrica Scandinavica* 125(6): 440–52.

Shoemaker, D. (2015). *Responsibility from the Margins*. New York: Oxford University Press.

Silberman, S. (2015). *Neurotribes: The Legacy of Autism and the Future of Neurodiversity*. New York: Avery.

Stegenga, J. (2018). *Medical Nihilism*. Oxford: Oxford University Press.

1
Quality of Will and (Some) Unusual Behavior

Nomy Arpaly

1.1 How I Will Not Use the Concept "Mental Disorder"

This chapter is about moral blameworthiness, or its absence, in unusual people who would normally be diagnosed as having well known (or somewhat less known) mental disorders, and I will, for convenience, refer to their conditions by the Diagnostic and Statistical Manual of Mental Disorders (DSM) categories that describe them best.[1] However, the concept of a mental disorder will not in itself play a role in my reasoning about their blameworthiness. An explanation of this absence is in order. I will provide it briefly here.

I do *not* in any way hold that the concept of mental disorder is only a "construction" whereby "society" controls deviant behavior, and it is important for me to emphasize that I reject this view. Watching a person with a simple phobia of spiders facing a spider is sometimes all one needs to be cured of one's youthful extremism on this topic. Perhaps some excessive reactions to arachnoids and insects are socially constructed—as part of femininity, say—but a true phobia does not seem like the sort of thing that can be "constructed" in the sense favored by comparative literature departments,[2] nor does one need to be prejudiced against the deviant, in the way posited by Thomas Szasz[3] in order to feel that there is something extraordinary about the arachnophobe that makes him suffer. When I say that philosophers need to be careful employing the concept of a mental disorder as used by contemporary psychiatrists I do not mean to defend a Szaszian position or engage in Critical Theory. However, the concept of mental

---

[1] American Psychiatric Association (2013).
[2] Continental suspicion of the concept of mental illness goes back to Foucault (2006).
[3] See especially Szasz (1961).

Nomy Arpaly, *Quality of Will and (Some) Unusual Behavior* In: *Agency in Mental Disorder: Philosophical Dimensions*. Edited by: Matt King & Joshua May, Oxford University Press. © Nomy Arpaly 2022.
DOI: 10.1093/oso/9780198868811.003.0002

disorder as used by psychiatrists today, and especially by the writers and users of the DSM, is not a philosophically respectable one.

I do not simply mean to say that "mental disorder" is not a natural kind. "Raptor" is not a natural kind, as hawks are not evolutionarily close to owls. Still "raptor" or "bird of prey" is in some contexts a legitimate theoretical category. "Mental disorder," on the other hand, is not a theoretical but a *practical* kind.[4] When trying to decide whether to define a person as having a mental disorder—is this person depressed or is he only grieving? Does this child have a mild form of autism or is she just a nerd?—practical considerations are brought in. For example, a psychiatrist might advocate for calling the grieving person "clinically depressed" and calling the socially awkward child "autistic" because if we were not to define the grieving person as clinically depressed, we will not be able to help them using medical insurance, or if we do not call the socially awkward girl "autistic," we will not be able to fund help for her in school. Another psychiatrist (or anti-psychiatrist) might object to defining either the bereaved person or the socially awkward child as having a mental disorder because it would be insulting to the man's grief if we called it a mental disorder or because it would stigmatize the already socially awkward girl to be declared to have a mental disorder.

These pros and cons of calling the conditions in question mental disorders are practical, not theoretical. Why should the fact that it is insulting to call a condition a disorder be theoretically relevant? Some people whose depression is severe and unrelated to any significant situation or event are gravely insulted when told that their condition might be caused by a glitch in the operation of their neurotransmitters and not by the deep insight into the world provided to them by having read Sartre. Still, their depression might well be exactly of the sort where we might suspect the more "medical" type of etiology. Similarly, why should the fact that a person needs help mean that the person is sick? A child who has no friends through K-12 and who is consistently beaten up by other children because they find her precocious taste for Shakespeare infuriating needs help, as suffering through K-12 with no friends and a lot of peer persecution is a horrible thing. That does not make a childhood taste for Shakespeare a *medical* matter, nor does it make distress over being shunned for one's taste for Shakespeare a disorder. *Not*

---

[4] For a detailed defense, from within philosophy of science, of similar sentiments see Tabb (2015).

all suffering is a medical matter and "medical" suffering is not the only kind of suffering that needs to be taken seriously.

I hope someday we'll articulate a respectable concept of mental disorder, or some other concept(s) that would enable good research into the relevant kind(s) of problems that people such as the arachnophobe have. However, as long as "mental disorder" is a practical kind rather than a theoretical kind, it is dangerous to use it in building a theory of responsibility. Thus, what I say about the agents I discuss is independent of whether or not "mental disorder" turns out to be a good expression by which to refer to their conditions.[5]

Let us now turn to my view of moral blameworthiness and see how it applies to ordinary behavior. We will then see how it can also be applied to cases of unusual behavior.

1.2 The Simple Quality of Will Theory

Consider the following case:

A Tempest in a Teacup: Ophelia and Amina are historians. Ophelia had sent Amina a message in which she asked her whether she happens to know a good article about Uriel da Costa's excommunication from the Jewish community in Amsterdam in the seventeenth century. To Amina, the story sounds familiar, and she has a sense that she had in the past skimmed an article on the subject, but seventeenth-century Amsterdam is not part of her specialty and she is unable to remember the title, the author or the publication venue. She decides to give herself a day or two to remember, leaving Ophelia's message unanswered. The trouble is that Amina is quite absent-minded and has a lot of work to do that week, and so she forgets about the matter completely. Shortly afterwards, Amina meets Ophelia at the history department, chatting with the department administrator, Gail. Ophelia expresses dismay at Amina for not answering her email. Embarrassed, Amina apologizes profusely for her absent-mindedness, saying that she forgot. Ophelia is unmoved. "Amina," she says, "maybe you forgot, but we know that if I were someone important or famous, you would have remembered." Gail then chimes in, telling Ophelia that she is simply mistaken. She says "Ophelia, there is no need to be so angry. As the secretary,

---

[5] That having a mental disorder per se does not have automatic implications for one's blameworthiness was argued by King and May (2018).

I can testify that Amina *always* forgets to answer emails. She forgets to answer emails from famous people. She forgets to answer emails from me about her tenure case. She loses checks made out to her and then forgets to ask me to reissue them. Don't take it personally."

Let us look at *A Tempest in a Teacup* more closely. All three characters assume that Amina had a (minor) moral duty to reply to Ophelia's email, if only to say "sorry but I don't recall." Ophelia clearly *blames* Amina for not answering her email and finds her *blameworthy* for failing to answer it. Gail, on the other hand, thinks Amina is *not blameworthy*, or is considerably less blameworthy than would warrant Ophelia's anger. The argument between Ophelia and Gail regarding Amina's blameworthiness or lack thereof centers on non-normative facts. Ophelia assumes that *if Amina cared enough*— about her duty to Ophelia or maybe about humans in general or politeness in general—she would have not failed to answer her message. Thus, Ophelia thinks that Amina's failure to reply to her email stemmed from indifference to Ophelia herself, to politeness, or to some other morally significant factor. Gail, the department administrator, holds that Amina, given her basic level of absent-mindedness, would have likely forgotten to answer the email no matter what, and so there is no reason whatsoever to think that her failure to answer is a manifestation of disregard for Ophelia, politeness or any other morally significant factor. Gail's evidence includes the fact that Amina, contrary to Ophelia's clear insinuations, is forgetful even when remembering is decisively in her own interest, and as most people care about their self-interest, it is reasonable to assume that if Amina's absent-mindedness does not lessen when her self-interest is on the line, her absent-mindedness is a cognitive problem that is not indicative of "not caring." Thus, Gail thinks that it is likely that Amina's failure to act, while it is wrong, is not the result of some moral indifference and therefore Amina is not blameworthy for it. In short, Ophelia and Gail both assume that Amina's blameworthiness or lack thereof depends on whether or not her course of (in)action stems from *lack of good will.*

The bare bones quality-of-will view of praiseworthiness and blameworthiness, one version of which is defended by Arpaly and Schroeder (2013) as Spare Conativism, is the view implicitly shared by Ophelia and Gail, only writ large. At the base of the theory is the idea that there are things that a moral person cares about, and caring about these things can be referred to, with a nod to Kant and a different nod to Strawson (1962), as "good will." When a person does the right thing out of good will, she is praiseworthy for the action. For an action to be blameworthy, it is not enough for it to be wrong, but it also needs to be manifestation of a shortage of good will—failure to care about the right things—or, alternatively, of ill will. A shortage of good will can be trivial, as when one can't be bothered to answer an email, or it can be dramatic, as when a person commits murder for money, indifferent to the moral status of the victim. Ill will happens when a person is motivated by considerations that are in essential conflict with the things that a moral person cares about—for example, if the moral person wants people not to suffer, a person who performs an action exactly because it would cause suffering thereby shows ill will. According to one somewhat less bare bones version of the view, Spare Conativism, there are also actions attributable to lack of *ill* will, but let's ignore that for now.

Timothy Schroeder and I have referred to lack of good will as "moral indifference," which is convenient, despite the awkwardness of discussing degrees of moral indifference, but it is important that this use of the word "indifference," as well as my use of the word "caring," is qualified in another way as well. It is natural English to say that my cats, Catullus and Philippa, do not care about or are indifferent to whether or not they damage my computer, as they do not possess the concepts "my" and "computer" or even a full-fledged concept of damage. This is not the kind of "not caring" or being indifferent to which I wish to refer. I am rather referring to the kind that is invoked when one asks one's spouse "do you not see the dust on the floor or do you *just not care*?" or when metaethicists wish to know if psychopaths do not know that what they are doing is immoral or know it full well but "just don't care." So, when I speak of "not caring" I am speaking of the colloquial "*just* not caring"—in other words, to cases where the agent can *conceive* the object of her indifference. That means that cats and babies cannot be morally praiseworthy or blameworthy: the cat doesn't conceive the fact that my computer is my property, and so she is not morally indifferent in our sense.

It is important to stress that the simple quality of will theory—henceforth the Quotidian View—is a view of the things that make a person blameworthy or praiseworthy for an action. It does not provide *diagnostic criteria* for blameworthiness or praiseworthiness because it is often very hard to tell whether or not an action is a manifestation of moral indifference (or good will, or ill will), mostly because we cannot read people's minds. Obviously, if a person fails to help you because he is tied to a chair, his inaction does not show moral indifference, but many cases are much harder to diagnose. Many people have wondered whether their spouses do not see the dust on the floor (in which case the fact that they don't clean does not show moral indifference) or, being messy themselves, do not see that someone might mind the dust on the floor (ditto) or just do not care enough about them to clean the floor (moral indifference). Many people have wondered if they themselves give to charity to help people (which sounds like good will) or to be perceived as good people (which does not) or maybe both. Many people have wondered whether the person who is rude to them is trying to upset them (ill will), does not care about their feeling (moral indifference) or is just socially incompetent or following habits from another culture (neither). Even if one could read people's minds, one would find hard cases involving mixed motives for action, motivated irrational belief, culpable ignorance, and other such complexities. Believing and wanting, on my view, are as different as oil and water, but in ordinary life they are very well emulsified.

In somewhat simpler cases, there exist some heuristics. One of the better ones is the one used by the department administrator in *Tempest in a Teacup*, which I would like to call the *self-interest heuristic*: asking yourself if the person in question behaves the same way when her self-interest is at stake. A person who would make a faux pas even when her self-interest is severely endangered by it is more likely to be socially incompetent or a cultural transplant than someone who only slights people that cannot harm her. As most people care about their self-interest, it stands to reason that if a person is absent-minded even when it comes to making sure checks are issued *for her* than her absent-mindedness in forgetting to return your email is no reason to suspect indifference: for all you know, your email is just as important to her as her interests are. Such heuristics, however, are far from perfect. The self-interest heuristic assumes that people care about their self-interest very much and are motivated to act for its sake, but these assumptions are not true of all people. The equal opportunity insulting person, for example, could be insulting her superiors because she does not care that much about her self-interest, or cares about it less than she does about some ideal of authenticity or contrarianism. But however hard it can be to answer questions like "if he cared, would he still have done it?" they are clearly questions that we often ask when we have trouble assigning moral credit or blame.

I will not attempt to offer a full defense of the Quotidian View or of any less bare-bones quality of-will account of moral blameworthiness, and I will especially avoid issues related to the classical problem of free will. What I would like to do is examine what the Quotidian View has to say about various kinds of unusual minds and behaviors of interest to psychiatry, psychology, and neuroscience. Essentially, this is an exercise in *parsimony*: can we explain the blameworthiness and praiseworthiness, or lack thereof, of people who are dramatically atypical in a variety of ways by appealing exclusively to the quotidian thing which is quality of will?

In the following sections, I'll take a look at some mental conditions and types of mental conditions and see what the Quotidian View has to say about them. I will not attempt to include all DSM categories, and I will specifically not discuss psychopaths and addicts, as I take the mission of this volume to be the discussion of conditions that are less discussed by ethicists and agency theorists.

1.3 Conditions Involving Epistemic Irrationality or Cognitive Impairment

Epistemologists as a rule take it be the case that to the extent that one fails to understand, say, topological set theory, one fails at rationality. I do not think this is true. I do not mean to say simply that the standard is high. I am perfectly willing to admit that not understanding topological set theory is a failure of smartness or a failure of intelligence—at least with regard to mathematics. But being smart is not the same as being rational and being unintelligent is not the same as being irrational. To see that, consider a normal child of 11 who is about as smart and as rational as 11 years old children generally are, and then consider the same child at 14. With brain development and with experience, the child will become smarter (ready for more difficult study material, able to learn new tasks with less help, etc.), but as many parents can tell you, it is likely that as an adolescent, flooded with strong desires, feelings and emotions, she will not be any more *rational* (or reasonable). Let me explain.

Roughly, the difference between a failure of intelligence and a failure of (epistemic) rationality is that a failure of intelligence involves an inability to acquire some concepts whereas a failure of rationality is what you suffer from if you fail to respond accurately to the relationships between concepts that you do, at the time, grasp. There are many concepts that my cats cannot acquire, but the fact that my cats cannot, for instance, grasp what philosophy is does not indicate that they are irrational. It indicates that they are *not that smart*. Irrationality is what happens when a person who seems to have a decent grasp on the concept "random" is nonetheless both under the impression that lottery tickets are chosen randomly and under the impression that this month's winning ticket is less likely to have the same number as last month's ticket. The most obvious cases of irrationality are cases involving desires or emotions, as when a person wishfully believes something against evidence even though, in matters on which he doesn't have particularly strong wishes, he is very good at responding to the same kind of evidence. "How did such a smart person make such an elementary error?" one might ask, and "wishful thinking" is a good answer, because wishful thinking isn't a failure of smartness, but of rationality.

This is not a work of epistemology, so I will skip some obvious complications and say that some conditions widely considered mental disorders for centuries—generally the ones that used to be called "insanity" or "neurosis"—involve *irrationality*, and many other medicalized conditions—most of which are thought of in terms of disability—involve *cognitive impairment* that does not in itself imply irrationality. Examples of irrationality include paranoid delusions, seeing a small spider as big because one is afraid of the spider, and the depressed person's tendency to see herself as honest-to-God terrible for doing things that, when other people do them, she regards as minor errors. Examples of cognitive impairment without irrationality are low intelligence, memory problems, learning disabilities, and inability to recognize people's emotions by looking at their faces.

It is easy to see that cognitive impairment can excuse from blame, and a bit harder to see how irrationality does so. Let us start from cognitive impairment. Cognitive impairment can excuse in two ways. First, if a person is so badly cognitively impaired that she cannot grasp morally relevant concepts like "harm" or "property" or "lie" she is in the same boat as my cats. She cannot be blamed when she causes suffering in other people, steals private property, or lies, respectively as the trouble with her is cognitive rather than conative or volitional. Again, this person cannot be accused of moral indifference, because moral indifference is being unmoved by morally important things of which one is aware, which requires the ability to grasp them. One class of people who might be in this extreme predicament of being unable to grasp morally important concepts are those who experience bad enough schizophrenic episodes that they speak, and seem to think, in "word salad."

A person who says, outside poetry, that she is "Germania and Helvetia of exclusively sweet butter"[6] probably does not express a belief—what would it

---

[6] As does an early twentieth-century patient Carl Jung mentions in his early work (Jung 1961).

mean for a person to believe she is Germany and Switzerland of exclusively sweet butter?—and it seems that her belief forming apparatus is severely damaged. This cognitive predicament is bad enough to exempt a person with moral responsibility, as the person does not seem to possess half-decent concepts of the things that a good-willed person wants or the things that an ill-willed person wants.

Second, in less severe cases, cognitive impairment can excuse through the ignorance that it causes. Factual ignorance often excuses, as many philosophers agree. To cite a famous example, a person who thinks (without irrationality, let us say) that she is putting sugar in another person's coffee cup but in fact is putting extremely sugar-like poison into the cup is not blameworthy for poisoning the coffee drinker, though in some cases she might be blameworthy for, say, keeping sugar-like poison on her kitchen shelves, which she could have expected to be confusing. A person who, because of low intelligence, thinks all white powder is sugar and who poisoned a person due to that alone—that is, *not* due to ill will or moral indifference—will not be blameworthy for putting white powder in a coffee cup, for the intuitive reason often phrased as "he didn't know," sometimes contrasted with "he didn't care." The same is true for the person who, due to dementia, manages to get confused between the sugar and the poison even though they were not placed next to each other, and puts the wrong substance in the coffee despite intentions.

This explains why people who are cognitively impaired are, when they are, excused from blame for their actions, but note that the Quotidian View does not imply that cognitively impaired persons are always exempt from blame. A person with low intelligence who can conceive of such things as suffering and property is excused from blame for actions that she does not understand—e.g., when she puts poison in coffee because she mistakes it for sugar. However, if she attacks a person violently, understanding full well that she is causing suffering and wishing to cause it, she can be blameworthy, even if she is unfit to stand trial. What is true of lack of intelligence is also true of types of cognitive impairment that only affect a relatively small domain of cognition. A person with autism is excused from blame when he hurts someone's feelings due to being bad at discerning feelings but *is* blameworthy for his action if he hurts someone's feeling on purpose. It is perfectly possible to be cognitively impaired *and* have ill will or be morally indifferent, as long as the cognitive impairment is not bad enough to make these attitudes impossible. The question in cases like these is always whether a particular action manifests ill will or moral indifference or whether it is *due* to the cognitive impairment.[7] As I have said earlier, it is sometimes hard to tell the difference in practice, and sometimes an action seems to need a hybrid assessment—for example, a child of ten might understand full well that by being violent towards a peer she causes him pain, but not be capable of anything like a proper idea of the long-term mental harm that peer persecution can cause a child who is regularly beaten up by other children.

Let us now discuss irrationality. As I have suggested, conditions that have been considered mental disorders for a very long time—depression, mania, psychosis, phobias—often involve gross irrationality, as it takes gross irrationality to believe one is Napoleon or to believe against clear evidence that one's family would be happy to see one dead. Gross irrationality often, but not always, excuses from blame.

People are very often quite irrational. How much epistemic irrationality is *gross* epistemic irrationality? I do not have the space here to formulate a full answer, but I will demonstrate what I have in mind by addressing an example: the difference between the average believer in astrology—an irrational-enough believer—and the person who has an honest-to-God delusion, such as the person with Delusional Disorder who believes that the FBI is after her.

Belief in astrology defies evidence as much as many delusions. Why, then, do we generally not take believers in astrology to be psychotic or deluded? A cynical answer is that contra Orwell, sanity can be a matter of statistics, and astrology is not taken to be a delusion because it is believed by a large number of people or by a sizable portion of the population. There might be something to this, but I think our failure to consider astrology fans deluded can be explained to a significant degree without such cynicism. The key factor in this explanation is the low credence most astrology fans have in astrological propositions—and here I include many people whose readiness to argue in favor of astrology till the cows come home might give you the mistaken impression that their credence in the main axioms of astrology is very high.

A common way to assess the credence one has in a proposition is the extent to which one would bet on it or "bank" on it, which is manifest in one's behavior and arguably, to some extent, in some of one's emotions, as one tends to be frightened or despairing if one has a high credence in a terrible proposition, happy if one believes in a happy one, and so on. Interestingly, how much one would bank on a belief does not correlate

---

[7] For an interesting discussion of cognitively impaired agents and their responsibility, see Shoemaker (2015).

with one's readiness to argue with others in defense of its truth or even fight a literal war against people who reject it—a belief to die for is not always a belief to bank on and vice versa. A person might be, in a way, very passionate about his belief in heaven and hell and still behave, and in some ways feel, like someone who neither hopes for heaven nor is afraid of hell. It would be expected that if one truly believed that committing one of the seven deadly sins might lead to hell, and that hell is a worse place than prison, one would avoid the deadly sins about as studiously as one avoids breaking laws the breaking of which would result in going to prison. It would be expected that if one believed that one will go to heaven upon death, and that heaven is a better place than Curaçao, one would find in the topic of death at least some of the cheerfulness that one finds in discussing an upcoming trip to Curaçao. Yet, many people who refer to themselves as believers are afraid of death and find it a relentlessly grim topic. Many such people are also as likely as many atheists to commit what they regard as deadly sins. They do not seem to *bank* on the existence of heaven and hell. In the words of a different Orwell character, they might believe in heaven and hell but they do not believe in them "the way they believe in Australia."

Most astrology fans do not bank on astrology. Quite literally, they do not use it to make critical investment decisions. While they might mention a desirable astrological sign in a personal ad, or use the excuse of being a Scorpio when being obnoxious to their partners, most of them will not get married or divorced for astrological reasons. In short, they find astrology a fun field and consider its opponents dogmatic, but they do not normally act on its advice the way they act on doctors' advice. This, in my view, is why they are not considered deluded. While it is counter-evidential and irrational to have even 10 percent credence in astrology, it is a lot less irrational than having 90 percent credence would be. A delusion, possibly the most irrational kind of belief, is not only a counter-evidential belief but a counterevidential *certainty*. Patients with delusional disorder who think that the FBI is after them do quit their jobs and run, and the delusion that someone is the devil can cause a person with schizophrenia to attack him so as to protect the earth, regardless of legal circumstances. They believe their respective falsities "the way they believe in Australia" and any beliefs that conflict with them are treated in the same way that a theory that denies the existence of Australia would be treated by you and me. I dare say that anyone whose belief in astrology was as firm as her belief in elementary geography and who used it to guide his action the way one uses a GPS *would* be described as deluded (or "crazy" or "nuts") by many.

What, then, is the connection between gross epistemic irrationality and blame? Here it seems useful to avail ourselves of a distinction made by David Pears (1984) between hot irrationality and cold irrationality. Hot irrationality is irrationality caused by emotion, desire, or some other motivation state (I mentioned wishful thinking). Cold irrationality is not caused in this way (I mentioned the gambler's fallacy).

Cold gross irrationality can exempt from moral blame in the same way that cognitive impairment or ordinary ignorance does. The person who attacks someone because she believes him to be the devil as a result of schizophrenia does not display ill will or moral indifference. Hot gross irrationality is more complicated. Imagine a person who believes—the way one believes in Australia—an elaborate conspiracy theory in which Jewish people play the role of super-villains. The content of the conspiracy theory is the product of "hot," motivated irrationality. Let us assume that our character hates Jewish people, and it's the hatred that inclines him to believe horrible and decisively counter-evidential things about them. To the extent that his irrationality is a symptom of such hatred—plausibly a form of ill will—it is hard to see the same irrationality as an excuse from blame. Thus it is natural to think of Hitler, assuming that he believed the views he expressed, as both quite irrational ("crazy") *and* evil. Of course, motivated and unmotivated factors can combine in making a person irrational. It might, for example, be "cold" neurological factors that determine whether your hatred of certain people will turn you into an ordinary conspiracy theorist or a psychotic one.

1.4 Quality of Will and Major Depression

Much depression involves or causes epistemic irrationality, much of which is fairly easy to detect. For example, if you are not clinically depressed, a dialogue between you and a (paradigmatic) clinically depressed person can look like the following:

TRISTAN: I am a terrible, horrible person and deserve to die.

YOU: Why do you think so?

TRISTAN: I forgot to buy milk, again.

YOU: My roommate always forgets to buy what she was going to buy in the grocery store, including milk. Does it make her a terrible person?

TRISTAN: No.

YOU: So forgetting to buy milk doesn't make you a terrible person.

TRISTAN: It does. I am different from your roommate.

YOU: How?

TRISTAN: In my case, forgetting milk is the result of basically rotten character.

A depressed person, then, can think that the fact that she forgot to buy milk makes good evidence that she is a terrible human being, whereas the fact that someone else—it often does not matter who—forgot to buy milk says nothing about his moral character. This is typical of the way the depressed mind processes evidence, heavily biased towards the depressed view of the world. The most extreme version of this kind of irrationality occurs when a person is convinced that her friends and family will be relieved if she commits suicide, where anyone else can see that they will be devastated. The moment where we suspect a person is no longer "just" sad because he lost his job but rather is clinically depressed is often the moment in which such irrationality occurs to a significant degree.

A depressive episode involving epistemic irrationality can excuse or partially excuse from blame like any other condition involving epistemic irrationality. For example, a colleague might miss a meeting in the midst of a depressive episode because of her high credence that her presence at the meeting would be useless or even harmful. This is not a manifestation of moral indifference. The self-interest heuristic often works here: depressed people miss meetings even when it is bad for them. The real challenge for the Quotidian View, however, is in cases of depression that do not seem to involve *any* epistemic symptoms.

"There are two types of depression, woe-is-me and what's-the-use," says a psychiatrist of my acquaintance, quickly adding that the same depressive episode can contain both the "woe is me" syndrome and the "what's the use" syndrome. Let us adopt this imprecise terminology for the moment. The "woe is me" patient's predicament is mostly epistemic, or at least cognitive: she believes she and the world around her are terrible, or at least ignores anything around her that might be good while paying attention to the bad. The "what's the use" patient's predicament is not epistemic: it is *motivational.* She no longer seems motivated to do things that she was motivated to do before, whether her motivation was prudence (e.g. she no longer pays her bills) or pleasure (e.g. she no longer bothers to visit friends).

As I have mentioned, the epistemic and motivational aspects of depression often appear together, and so can be hard to tell apart. A person might "not bother" to visit friends in part because she no longer feels loved by her friends and is inclined to believe that they hate her, and as a result no longer enjoys their company. A person might believe, as part of her epistemic irrationality, that she is an incompetent worker, and thus lose motivation to do anything complicated at work (or to go to work at all). But imagine— or recall—a person whose depressive episode could be described as pure "what's-the-use," except that strictly speaking, he does not even have the *belief* that there is no use doing anything. His belief-forming apparatus seems to be fine, but he is unmotivated to an extreme degree. Suppose such a person—call him Seth—does not appear at a meeting at work. Suppose there are moral reasons to be at the meeting. Intuition, for many of us, says that he is less blameworthy than a typical person who skips the meeting for no compelling reason, and the self-interest heuristic encourages this line of thinking, as Seth, too, is likely to miss meetings that are in his self-interest to attend. In what way is he not displaying lack of good will, aka moral indifference? Or, to put it more colloquially: *why do we not just say that he is lazy?*

A lazy person, let us assume, would miss the meeting because they prefer to do something else that's easier—say, watch TV. Seth has a different motivational (or de-motivational) story. My suggestion is that though he does not believe that "it's no use" going to the meeting—that nothing good is to come out of going, morally or prudentially—he has the gut-level expectation that it would all for naught, that all his actions will fail to do him, or anyone else, any good.

What are gut-level expectations?[8] I am not referring here to beliefs about the future that are unreflective or unconscious. Such beliefs can in fact influence our behavior through the guise of "hunches," "instincts," and so on, but these are not the mental states I am discussing here, but rather things that are not, strictly speaking, beliefs at all, though they fit within the related and broad philosophical category of "a-lief."[9] Consider a person who is too afraid to step onto a transparent bridge high above the ground but has no qualms about allowing her children to explore it cheerfully. That person— let's call her Fatma—clearly does not believe, at any level, that she will fall to the ground below if she steps on the bridge. Even an inarticulate or

---

[8] For a full answer based on empirical data see Schroeder (2004), chapter 2.
[9] Introduced by Gendler (2008). I do not simply use Gendler's term because I am not convinced that none of the states classified as a-liefs, especially given the post-Gendler literature, are in fact beliefs of an inarticulate or unconscious sort.

unconscious suspicion that the bridge is unreliable would have caused her to be afraid of allowing her children to step onto the bridge, but Fatma just smiles and says, with a slight embarrassment, "I guess they are braver than I am." However, looking down through the glass invokes a stubborn visceral expectation of falling, and so she is too terrified to step onto it.

A common context in which we run into gut-level expectations that don't seem to be beliefs is the context of what is known as "getting used" to things. Consider the following case:

It is 25 degrees Fahrenheit in Spencer's town and has been around this temperature for a while. Spencer flies to Florida. When he gets off the airplane, he is overcome by the pleasant warmth and experiences joy that the locals who come to meet him do not experience on that occasion. That is because there is a sense in Spencer's body "expects" a much colder temperature, and the actual temperature feels so high by comparison. Spencer need not, consciously or unconsciously, believe that he is going to be cold when he gets of the plane. In fact, Spencer might be thinking nothing but "Florida, here I come" throughout his flight, excitedly anticipating—that is, *cognitively* expecting—the sunny weather. However, he is used to—that is, he viscerally expects—a lower temperature, and enjoys the contrast.

Here is another case:

Paula always did well in school. She was looked at as gifted and talented starting in kindergarten. She got excellent grades in primary school and in high school. She got excellent grades in college. She did not always get the best grades in her class, but her grades were always excellent. When she applied for admission to graduate school in philosophy she got into one of her top choices. There, in graduate school, she was warned many times by her teachers that the job market in philosophy is very harsh and that even the best students are not unlikely to find themselves without jobs, or even without interviews. She received reliable statistics and did her best to brace herself for the possibility of not getting a job. She had no illusions about the quality of her work—if anything, like many graduate students, she had become insecure about its quality. Still, when her first attempt at getting a job in philosophy results in two interviews and no job offer she is dealt a brutal emotional blow that would seem more congruent with a *surprising* misfortune than an expected one. That is because Paula is *viscerally* surprised. She had gotten *used* to things going well for her when it comes to anything to do with academics.

My suggestion is the following: the victim of the pure episode of what's-the-use depression viscerally expects, and strongly so, that all his actions will not lead to any improvement in his state or in the state of the world. Like the fear of falling can be for some people who cross a glass bridge, the visceral expectation is very powerful for Seth, and he approaches every task in life with the sense of resignation in which one would approach finding a needle in a haystack. That, rather than any preference for lying in bed, is what keeps him from the meeting, and that earns him at least a partial excuse, depending on the severity of the depression.

Arguably, gut-level expectations seem to be, in general, a part of what it means to be in a certain mood or what typically results from being in a certain mood. A typical person who is in a good mood because of just having been to an enjoyable concert might be, as a result, more viscerally optimistic than before about the results of the elections taking place the next day. It need not be the case that the concert experience changed her *beliefs* about politics. A person who is manic but not psychotic might viscerally expect things to go her way when it comes to investing in the stock market even if her beliefs about the stock market haven't changed much. Such exploration will have to wait for another day.

My proposal regarding depression, in addition to explaining the difference between the depressed and the garden variety lazy TV aficionado, also distinguishes the real-life depressed person from the "depressed" person as described by philosophers when they need an example of putative moral belief without motivation. That imaginary person is described as genuinely having *ceased to care* about morality (or about the morally important things *de re*) and remained only with causally inert beliefs about what she ought to do. An amoralist with moral beliefs would make a fascinating case, but has little to do with the depressed person next door. It might have more to do with people who, due to injury, become psychopaths late in life, apparently without having lost any beliefs.

1.5 Non-excusing Psychiatric Predicaments?

The person diagnosed with Factitious Disorder either pretends to be sick or intentionally produces real sickness in herself in order to receive positive attention from the people around her. She is different from the "malingerer," the person who fakes illness for a more tangible benefit like avoiding the need to work for a living or avoiding military service. The "factitious" patient would in fact work harder than necessary at her job and refuse the help she is offered, because in this way she will receive admiration in addition to compassion. It is a good question why pretending to have cancer in order to avoid work is not considered a mental disorder, but pretending one has cancer in order to evoke compassion and admiration is. The fact that the latter is stranger, or even the fact, if it is a fact, that it is likelier to harm the agent, does not seem to be enough of an in-principle reason.

From the Quotidian View, there is no prima facie reason to regard the "factitious" patient as less blameworthy for her deceptive behavior than a pretender who does not qualify for a DSM diagnosis. In fact, I suspect that by Quotidian View standards, there are some malingerers who are less blameworthy than some factitious patients. For completeness, I'll mention that there are malingerers who avoid military service for good moral reasons, and though they are not as brave as conscientious objectors, they are praiseworthy. Even if we restrict ourselves to malingerers who act in their own self-interest, some such malingerers act to avoid a truly terrifying prospect, in which case their actions do not show ill will or serious moral indifference—whereas a factitious patient, as traditionally described, might be merely seeking to remedy a lack of sympathy in her life.

What if a factitious patient does not *merely* crave sympathy the way most people do, but suffers from unusually intense self-hatred or an unusually shaky sense of self-worth? That would make her equal in her suffering to some people who are clinically depressed. A person who acts immorally to counteract serious depressive symptoms is not, per the Quotidian View, as blameworthy as a person who acts immorally in order to get rich—it doesn't take as high a degree of moral indifference to be tempted by the avoidance of pain. It might also be true that it makes sense to feel compassion for such a person despite her blameworthiness—and the Quotidian View is not in any way committed to the thesis that blameworthiness always entails punishment being right or compassion being out of place. Still, even if the factitious person is depressed, she might compare badly to many people who are depressed—mildly, moderately, or severely—and do not deceive or manipulate anyone. After all, many depressed people are *overly* concerned about being a burden on their friends and family while the factitious patient makes herself considerably more of a burden when she pretends to have, say, cancer. Given the way factitious disorder is described in the DSM,

I suspect that the Quotidian View might need to bite the bullet and say of some of these patients that they are not excused from blame, as well as say of some others that they are only partially excused from blame. Details do vary, though, and sometimes depression has the power to defeat even a very good will.

One of the most frightening persons in the DSM, second only to those with psychopathy, is the person diagnosable with the form of Factitious Disorder known until recently as "Munchausen Syndrome by Proxy"[10]—a person who induces serious illness in her child or who drags her child through painful and dangerous procedures on the basis of symptoms she pretends the child has. If the story as currently told by psychiatrists is true and the person in question is simply motivated by a desire for compassion and admiration, it is hard not to see her as a case of chilling moral indifference, an uncaring person willing to make a child suffer for the sake of "playing the martyr."

Another of the more frightening people in the diagnostic manual of mental disorders is the narcissist, and he, too, might be a case of moral indifference—at least if the story often told about his psyche is true. A popular theory is that the narcissist "overcompensates" for profound insecurity through his self-centeredness. If narcissism is in fact a way of dealing with insecurity it is, again, hard not to see the narcissist as a selfish person—in the ordinary sense of someone who prioritizes her wellbeing over that of others more than a half-decent person does. Severe insecurity is unpleasant, and can be an *extenuating* circumstance for some actions, but there is a limit to how much one can deal with emotional displeasure at the expense of others without counting as a case of significant moral indifference. Again, it might be that the genesis of narcissistic behavior is different and has nothing to do with insecurity. If scientists discover such a genesis the verdict of the Quotidian View might have to change along with the story.

It should be added that some DSM categories are so broad, roughly defined, or, one suspects, applied so liberally that I expect each of the relevant diagnoses is given to some people who are blameworthy to various degrees for their characteristically bothersome behavior and some who are blameless. This controversial territory will have to be covered another day.

As I have warned, these have only been very few of the mental conditions discussed in the ever-expanding DSM. Most clearly missing in this work are

---

[10] Today it is officially known as Factitious Disorder Imposed on Another.

conditions that involve seemingly uncontrolled impulses, compulsions, tics, and other forms of unusual motivation. Arpaly and Schroeder have already discussed addiction (2013) and Schroeder has discussed Tourette Syndrome (2005) from an angle sympathetic to the Quotidian View, but I hope to be able to discuss other types of compulsion-like urges and impulse control issues in future work. Meanwhile, I hope I have given you a decent idea of a way the same quotidian intuitions that guide us when judging people's more boring actions can be stretched to help us with more interesting agents, even the sort whose conditions will be prime candidates for the category "mental disorder" when it becomes more philosophically respectable.

References

American Psychiatric Association. 2013. *Diagnostic and Statistical Manual of Mental Disorders* (5th ed.). Arlington, VA: Author.

Arpaly, N. and Schroeder, T. 2013. *In Praise of Desire*. New York: Oxford University Press.

Foucault, Michel. 2006. *History of Madness*. J. Khalfa, editor, translator, and J. Murphy, translator. New York: Routledge.

Gendler, T. 2008. "Alief and Belief." *The Journal of Philosophy* 105(10), 634–63.

Jung, C. G. 1961. *Collected Works, Vol. 3: The Psychogenesis of Mental Disease*. Princeton, NJ: Princeton University Press.

King, M. and May, J. 2018. "Responsibility and Mental Illness: A Call for Nuance." *Neuroethics* 11 (1), 11–22.

Pears, D. 1984. *Motivated Irrationality*. Oxford: Clarendon Press.

Schroeder, T. 2004. *Three Faces of Desire*. New York: Oxford University Press.

Schroeder, T. 2005. "Moral Responsibility and Tourette's Syndrome." *Philosophy and Phenomenological Research* 71 (1): 106–23.

Shoemaker, D. 2015. *Responsibility from the Margins*. New York: Oxford University Press.

Strawson, P. 1962. "Freedom and Resentment." *Proceedings of the British Academy* 48: 1–25.

Szasz, T. 1961. *The Myth of Mental Illness*. New York: Harper and Row.

Tabb, K. 2015. "Psychiatric Progress and the Assumption of Diagnostic Discrimination." *Philosophy of Science* 82 (5): 1047–58.

2
Disordered, Disabled, Disregarded, Dismissed

The Moral Costs of Exemptions from Accountability

David Shoemaker

Whatever else you think of his view, there remains something quite attractive about P.F. Strawson's core assumption that the capacities for being a
morally accountable agent are just the capacities for being in "ordinary adult
human relationships..." (Strawson 1962/2003: 81). There is indeed something compelling about the idea that accountability is *to* others, others with
whom one not coincidentally also stands in various relationships, so that
what it takes to be in those relationships with others is just to be susceptible
to being held to account by them for failing to adhere to the norms and
expectations that define those relationships as such. Being *excluded* from
interpersonal life, then, is just to be *exempted* from accountability, and vice
versa. In ordinary interpersonal life, this wholly overlapping "exclusion-
exemption" is most often illustrated by the treatment of people with serious
psychological disorders.

   Now when people are excluded from valuable domains on the basis of
their arbitrary characteristics (such as race and sex), they are discriminated
against, prevented from receiving the benefits of participation in those
domains for morally irrelevant reasons. Accountability is also such a
domain. Exemption from it—via exclusion from the interpersonal
domain—thus seems to prevent exempted parties from receiving crucial
human goods for morally irrelevant reasons. Exemption thus seems a form
of morally objectionable discrimination against those viewed as having what
I will label *accountability-disabilities*.

   In this chapter, I will discuss two widely deployed ways of trying to
ameliorate morally costly disabilities. Both fail to be viably applicable to
accountability-disabilities, however. I will thus sketch my own solution to

David Shoemaker, *Disordered, Disabled, Disregarded, Dismissed: The Moral Costs of Exemptions from
Accountability* In: *Agency in Mental Disorder: Philosophical Dimensions.* Edited by: Matt King & Joshua May,
Oxford University Press. © David Shoemaker 2022. DOI: 10.1093/oso/9780198868811.003.0003

the problem, one that involves disentangling accountability and interpersonality in a way that also provides insights into our shared human nature.

2.1 Accountability and Interpersonality

Strawson introduced the core assumption, but many theorists have since adopted and assumed it as well, including those who disagree with many other aspects of the Strawsonian approach.[1] On Strawson's original view, what it means to be a responsible agent is just for one to be regarded as an appropriate target of a set of (mostly) emotional responses, what he called "reactive attitudes," such as resentment, indignation, and guilt. These are our natural responses, he claimed, to violations of our standing demand for good will on the part of others. They are the ways in which we hold people to account for those violations (thus the standard contemporary label I have adopted for the kind of responsibility that's at issue here, even though Strawson himself doesn't use it: *accountability*).

The reactive attitudes are what Strawson also termed "participant" attitudes (which also include gratitude, forgiveness, love, and hurt feelings), and it's the susceptibility to them, he thought, that is constitutive of "involvement or participation with others in inter-personal human relationships . . ." (Strawson 1962/2003: 79). As Gary Watson interprets Strawson, "[O]ur social sentimental nature grounds the distinctive reasons that structure our personal relations"—we care about how these others regard us—and so gives rise to the *basic demand* "to be treated with regard and good will" (Watson 2014: 17). Thus "[t]o be [an accountable] agent is to be someone whom it makes sense to subject to such a demand" (Watson 2014: 17).

It makes sense to subject people to this demand—to treat them as accountable—only if they are able to understand and speak the emotional language people use to react to its violation. But insofar as this is just the language of interpersonal relationships, the membership conditions for the two domains—accountability and interpersonal relationships—are identical. People thus ought to withhold the reactive attitudes of accountability generally from those seen as "*incapacitated* in *some or all* respects for ordinary

---

[1] See Stern 1974; Watson 2004: 219–88, and 2015; McKenna 2012; Darwall 2006; Shoemaker 2007 and 2015. Others, while perhaps not viewing the capacities as identical, at least tie them very closely together (see, e.g., Bennett 1980; Fischer and Ravizza 1998: 208–14; Scanlon 2008; Russell 2013).

inter-personal relationships" (Strawson 1962/2003: 82; first emphasis in original, second emphasis mine).

On Strawson's view, what we ought to take up instead toward incapacitated agents—if we are "civilized" (Strawson 1962/2003: 81)—is the *objective* attitude, viewing them as objects "of social policy; as [subjects] for... treatment; as something certainly to be taken account... of; to be managed or handled or cured or trained..." (Strawson 1962/2003: 79). And this is, indeed, how many people view those with severe psychological disorders, as neither accountable nor as those with whom one can be in a genuine interpersonal relationship.

The difference here is one of default emotional stances. People stand ready to engage emotionally with most others via the entire range of reactive attitudes, that is, the default stance people have toward others is the emotionally engaged participant stance, where they are ready to respond with resentment or gratitude, say, depending on what the others do. But once certain bits of information are revealed about *some* possible targets, people's default stance tends to switch to a readiness *not* to engage emotionally with them via the reactive attitudes, and they take up an emotionally reserved objective stance instead. This latter emotional default stance is all that's meant by talk of "exempting" people from accountability.

And who are the people regularly exempted from accountability? Strawson gives just a few examples: "hopeless schizophrenic[s]," those whose minds have been "systematically perverted," those who are "warped or deranged, neurotic or just a child," as well as those, finally, who are "compulsive in behavior or peculiarly unfortunate in... formative circumstances" (Strawson 1962/2003: 78–9). These are all, importantly, people who are *globally* exempt, people incapacitated, as he says in "all respects for ordinary inter-personal life," and so people toward whom folks tend to suspend their entire set of participant attitudes in every domain. Non-disordered folks can't ever hold them accountable because they can't ever be in interpersonal relationships with them.

Now while Strawson does mention that what he says also includes those agents incapacitated for only "some" aspects of interpersonal life, he doesn't give any examples. But these *locally* exempt agents are very familiar. They include people with autism, who are often treated objectively—excluded and exempted—just in the social domains that seem to require the capacity to read people's intentions off of their behavior and facial cues. Those with clinical depression are often excluded and exempted just in interpersonal domains where certain motivational strength is demanded. And psychopaths may be excluded and exempted just in those moral domains requiring the capacity to recognize and respond to reasons grounded in others' interests.

Strawson was simply trying to articulate what he took to be the complete overlap of the conditions of exclusion and exemption already widespread in our ordinary practices, and it is indeed quite a familiar phenomenon (which likely explains why several non-Strawsonians buy into the idea as well). When people come to find out that someone is clinically depressed, for example, they tend to drop their ordinary expectations and readiness to engage emotionally with her when it comes to the domains in which her disorder has its greatest influence (e.g., motivations and affect). And generally, when people find out that someone has a "mental illness," they tend to recoil, fearful, and tend to avoid socializing with, falling in love with, or hiring that person as a babysitter, for example (see, e.g., Rabkin 1974; Bhugra 1989).

In this chapter, I will be talking mostly about local exemptions. Very few agents are so systematically impaired as to be globally exempt. Local exemptions are taken to apply to lots of people, though, not only those already mentioned, but also those with various degrees of dementia, dissociative identity disorder, anti-social personality disorder, obsessive-compulsive disorder, post-traumatic stress disorder, or various eating disorders. It is these "psychologically disordered" people who are most often exempted from some arenas of accountability, who are treated as being *disabled* for it. We thus have a familiar and widespread practice—which Strawson merely identified, distilled, and articulated—in which people with various psychological disorders are treated as exempt from accountability for the very same reasons they are excluded from interpersonal life. But there are huge moral costs to this practice.

2.2 The Moral Price of Exemptions

Interpersonal relationships are constituted by mutual demands, expectations, and exchanges of good will, affection, esteem, fellow-feeling, friendliness, and, most importantly, recognition and regard. Excluding people from this domain, and so exempting them from the reactive attitudes constitutively attached to it, is quite morally costly. I will focus here on two significant moral costs.[2] First, exclusion-exemption cuts one off from fellow-feeling and emotional engagement. This is the domain of friendship and love, after all. But people report how their caring and loving attitudes tend to dissipate toward spouses who have Alzheimer's dementia (see, e.g., Hayes et al. 2009), or how tough it is to remain emotionally open and vulnerable to friends and family members with PTSD (Matsakis 2014) or traumatic brain injury (Nabors et al. 2002). But emotional withdrawal inevitably affects those withdrawn from. As Jonathan Glover puts it, "[T]o withhold the reactive attitudes is to exclude those individuals from a central part of human relationships," which "seems unfair" (Glover 2014: 304; emphasis added). Exemptions involve *emotional starvation*.

Second, excluded-exempted agents are *denied recognition and regard*. The basic demand for regard is presumed to be mutual: My demand that you recognize my worth and regard me appropriately is just the correlate of my expectation that you demand the exact same thing of me. To be outside of this reciprocal relation is to be outside of the domain of recognition and regard. But as Jonathan Glover has powerfully shown, there are serious moral costs to being banished from this domain. In probing conversations with a large number of violent patients with Anti-Social Personality Disorder (ASPD) in Broadmoor psychiatric ward about their values, Glover found that they tended to care about only very superficial ways of being and living, and that they had only what he called "a weak sense of moral identity" (Glover 2014: 56). But there were also several common themes of their childhoods that likely contributed to their having such shallow moral identities: severe abuse, humiliation, guilt-inducements, self-hatred, a lack of control, and, most importantly, serious disrespect and lack of recognition from others (Glover 2014: Chs. 3 and 26). If lack of recognition and regard is precisely what sometimes contributes to the development and maintenance of some mental disorders like ASPD, then the costs of exclusion-exemption are significant indeed.

Given that exclusion-exemption deprives many of our fellows these two significant moral goods—despite Strawson's very British assertion that doing so is "civilized"—these agents seem subject to an objectionable pattern

---

[2] There are surely other costs, including being patronizingly excluded from demands for basic good will, suffering epistemic injustice for not being included amongst the community of reason-exchangers, and being denied the prudential benefits associated with being seen as an enforcer of social norms (thanks to August Gorman and Shaun Nichols for discussion of these last two). Because these moral costs are downstream and derivative from the two major costs I discuss in the text, I set them aside here.

of systemic *discrimination*. They are being treated as subordinate—not worthy of goods like fellow-feeling, recognition, or regard—solely on the basis of their psychological disorders. These disorders are thus being viewed and treated as *disabling*, for both interpersonal relationships and accountability.

When it comes to ameliorating discrimination against people viewed and treated as having disabling features, there have been two general strategies, differentiated in terms of how "disability" is to be properly modeled: (1) a "disability" is a tragic physical or psychological flaw in individual agents that it's up to medical experts to treat and fix, so as to render them newly able to access the goods and opportunities of which they have been deprived (the *medical model*); or (b) a "disability" is a socially constructed category, constituted by prejudice and discrimination against people for various physical or psychological differences, and so disability is *society's* problem, the solution to which is *accommodation*, the elimination of socially con-structed barriers preventing people's access to goods and opportunities (the *social model*). In the next two sections, I will explore what each model might say about the excluded-exempt. I will do so by focusing on the "accountability-disabled." To foreshadow a bit, in the literature on account-ability for "disabled" (non-paradigmatic) agents, theorists almost invariably adopt a kind of medical model of disability. By contrast, in disability studies, the medical model has long been discarded in favor of the social model, but there has been no accompanying discussion of what doing so means for those who are "disabled" for *accountability*. My aim is to see what might be gained by bringing these two literatures together. As we will see, neither can adequately ameliorate "accountability-disability" on its own. A new approach is thus needed.

2.3 Applying the Medical Model: The Project of Understanding

Jonathan Glover, a philosopher of psychiatry, offers the clearest and most charitable deployment of a kind of medical model construal of those treated as exempt from accountability in virtue of their psychological disorders. He favors a kind of "deep self" view of accountability, and so his aim is to discover what people's *moral identities* are, as well as the extent to which those moral identities determine their actions and attitudes. To have a moral identity is to care about the kind of person one is and wants to be, an ideal awash in deep(er) values (Glover 2014: 53–61). So to the extent that one is generally able to manifest one's own true values in one's actions and attitudes, one is an accountable agent. And to the extent that one's actions and attitudes tend to depend on something *else*, like a psychological disorder, one is exempt from the domain of accountable agency.

Glover's guiding aim is ultimately to figure out how to help disordered people get better. It often seems that there is a wide gulf between the non-disordered and the disordered. The key motivation of Glover's work, then, is to bridge that gulf by coming to *understand* the disordered, that is, to figure out what their moral identities are and the extent to which those identities do or don't determine their actions. This involves finding a plausible interpretation of what they are doing and why they are doing it, one "that can help break down the isolation, the 'gulf beyond description'" (Glover 2014: 127). We are to look first and foremost for the causal explanation of their behavior. Once we find it (is it the person or the disorder?), we'll have greater insight into how better to treat them, psychiatrically, so as to eliminate their accountability-disabilities. Call this the *Project of Understanding*.

Obviously, the extent to which we tend to be able to understand other people depends most heavily on how similar we both are. The more different other people are, the harder understanding them is. And some people with psychological disorders may seem particularly hard to understand, thinking and behaving as they do in seemingly bizarre ways. Recognizing this difficulty, Glover attempts to lead by example, devoting most of his book to showing how to gain understanding of many different disorders, and providing insight into how those with the disorders might be treated and *turned into* accountable agents.

I can only discuss a few of his many valuable case studies here. Start with ASPD. The low degree of empathy and the shallowness of the moral identities of those with ASPD suggest that they are less than accountable agents in many moral and prudential domains. They had "constraints on self-creation" imposed on them by their typically horrific childhoods (Glover 2014: 304).[3] To enable their accountability status, then, what's needed is help in "building up a coherent moral identity, a sense of who they are that will enable them to live outside in the world and to live at peace with themselves" (Glover 2014: 74). Crucial to this task is providing them

---

[3] However, Glover also signals a kind of ambivalence toward such people, as they do have "appalling attitudes toward [their] victims" (Glover 2014: 304).

with the kind of recognition and respect that was absent from their early lives, the need for which "found expression in violence" (Glover 2014: 43).

Glover interprets those in the grip of a serious eating disorder like anorexia nervosa, on the other hand, as having values and a moral identity already: What seems of overwhelming importance to them is maintaining a certain weight and body shape. These are typically values that they didn't used to have, however (or that weren't nearly as dominant). He suggests that those values "seem to reflect not the person but the trap they have fallen into" (Glover 2014: 355). If he is right that the moral identities in play are not really *theirs* while in the grip of the illness, then the actions and attitudes manifested by the illness-caused values are not theirs either, that is, they are exempt from accountability. To help them become accountable, a therapist should help them "engage in dialogue that may help them decide which values to make their own" (Glover 2014: 365).

The verdict is mixed regarding those with schizophrenia. Sometimes the "alien voices" in their heads generate behavior that others cannot make sense of as anything but a product of their illness. But there remains the possibility of "reclamation" for those with schizophrenia. And indeed, from the inside, things may seem quite different. While they may at first fight against the disease, they may also eventually come to terms with it and actually come to *integrate* it into their self-conception, ultimately establishing a new and different moral identity (Glover 2014: 385–6). In such cases, psychiatrists may engage in treatment to help their patients restore both their autonomy and their accountability status. This involves understanding their passivity in the face of the symptoms, but also encouraging them to come to view themselves in a different, more active, way (Glover 2014: 387).

With real effort, then, the non-disordered may be able to come to understand the disordered, and so come to determine whether or not they are to be included in the domain of accountability. If they are, great; if they aren't, insight will have been achieved about how best to treat them and perhaps eventually render them capable of accountable agency.

Again, this is to view those with accountability-disabilities through the lens of the medical model of disability. The major problem with this model, however, is that it completely overlooks the role of environmental and social conditions in *constituting* disability. Instead, it looks at "disabled" people exclusively in a negative individualistic light, as flawed agents, disempowered and living poor lives, rather than seeing the political, environment, and social conditions that are behind the discrimination and exclusion against them. This is why those from disability studies have mostly rejected it.

Now to be fair, Glover's own approach certainly doesn't commit these more egregious sins, nor does he entirely overlook the environmental contributions to psychological disorders. Indeed, he brings to our attention precisely how horrifying childhoods are among the causes for developing ASPD. But that people have been *caused* by their environments or society to develop "disabling" features doesn't yet acknowledge how environmental, social, and political factors have rendered those features *"disabling"* in the first place. Instead, Glover continues to treat the psychologically disordered as being "disabled" in virtue of their individual agential incapacities. And this is what has seemed to many in the disability studies movement to be a serious mistake.

Glover's Project of Understanding, and a medical model of disability generally, is unsatisfactory with respect to our motivating moral problem. Perhaps instead of trying to change disordered agents to eliminate their accountability-disabilities, therefore, non-disordered agents should try changing *themselves* in order to better accommodate those with disorders into the accountability community as they are. Indeed, this is the main recommendation coming out of the leading work in disability studies.

2.4 Applying the Social Model: The Project of Identification

Most people have jettisoned the medical model of disability in favor of a *social model*, according to which "disability" is a socially constructed category. There are two crucial elements of the model: *impairments* and *social prejudice*. Disability is then taken to be "entirely constituted by social prejudice against persons with impairments" (Barnes 2016: 25; for additional articulations of the social model, see Oliver 1996 and Oliver and Barnes 2012). Consequently, if society (and interpersonal norms) were designed differently, "there would be no disabled people" (Barnes 2016: 25).

While it has its critics (see, e.g., Wolff 2011: 163–4; Shakespeare 2013; and Barnes 2016: 27), the social model has made a significant dent in the public consciousness. Consider, for example, the provision of access ramps and wider bathroom stalls for those in wheelchairs, internet and telephone access for those with hearing, visual, and/or speech impairments, closed captioning for TV broadcasts, and much more. The idea is powerful: Those with "disabilities" have been historically disenfranchised for their mere differences, prevented by the ways people have designed and constructed buildings, say, from taking advantage of or equally competing for numerous opportunities for which they are perfectly well qualified. Viewing the "disabled" through the lens of the social model, then, yields an obvious normative recommendation: If the social environment can be altered so that their physical and cognitive differences no longer prevent them from taking advantage of the available opportunities, then their full equality as citizens can be established and buttressed, something there is surely powerful moral reason to bring about.

Can we thus view those with *accountability*-disabilities through the lens of the social model? As I noted earlier, there aren't any examples of this method for us to follow in the responsibility literature, as the overwhelmingly dominant perspective taken toward exempted agents has been through the lens of the medical model.[4] For standard responsibility theorists, viewing exempted agents through the lens of the social model would require a dramatic paradigm shift. It would require viewing "exemption" as a socially constructed category, consisting of "psychological disorders" *plus* "social prejudice against people with those disorders." Insofar as we have powerful moral reason to eliminate discrimination, then, we ought to engage in the kind(s) of self- and social-reconstruction that would accommodate people with these disorders within the accountability community *as they are*, thus making their accountability-disabilities disappear.

We don't have a Glover here to help us see how to do this, but I think we can pull together and draw from two different literatures to figure out how it is supposed to work. I will focus here on *dyslexia*. Michael McKenna gives us a great example of what it looks like from the philosophical literature when we see psychological disorders and exemptions through the lens of the medical model:

> In years past, the child who was a poor reader was often scolded for her poor performances ("Lazy child, she should just try harder!"). But we have

---

[4] There may seem to be a few nearby exceptions. One is Sneddon (2005), who argues that social conditions and context matter for the determination of one's status or degree of responsibility. Another might be Sommers (2012), who points to differences in moral ecologies to explain differences in responsibility statuses. And work by Vargas (2013), Fricker (2016), and McGeer (2019), discusses ecological contributions to responsibility impairments that may be ameliorated by addressing those social and environmental conditions (to build, as Vargas puts it, "better beings"). None of these writers are explicitly treating the "responsibility-impaired" as *disabled*, though, and regarding Vargas, Fricker, and McGeer, their thoughts about changing the responsibility-ecology are really aimed at providing the "responsibility-impaired" with the tools necessary to *become* responsible agents, which is just another application of the medical model in the end.

since learned that with some children dyslexia impedes the natural learning process so that mere exertion of added effort is ineffective. So we
revised our [accountability] practices accordingly....

(McKenna 2012: 50)[5]

That is to say, people have come to exempt those with dyslexia—switched
their default emotional stances toward them—when it comes to appraisals of
accountable effort in the reading domain, as it is thought that dyslexics
simply lack sufficient psychological capacities to conform to otherwise age-
appropriate reading demands.[6] As they are, people with dyslexia have a
reading-disability, and until they can be successfully treated, that disability
will remain.

Social modelers treat dyslexia in a starkly different way. As Elliott and
Gibbs (2008) tell the story:

There appears to be no clear-cut scientific basis for differential diagnosis of
dyslexia versus poor reader versus reader. At various times and for various
reasons it has been a social convenience to label some people as dyslexic
[for purposes of school assessment and funding, e.g.,] but consequences of
the labelling include stigma, disenfranchisement and inequitable use of
resources....Proper treatment is...hindered by the false dichotomy
between dyslexia and non-dyslexia. Let's not ask, 'Does dyslexia exist?'
Let's instead concentrate upon ensuring that all children with literacy
difficulties are served.   (Elliott and Gibbs 2008: 488)

So how do we do so? Recognize, first, that students learn things in different
ways, so identify what these are and then reconstruct the social environment

---

[5] I should emphasize that McKenna is not by any means advocating a simplistic medical
model himself. Indeed, he has the resources in his account to make some of the moves I suggest
a social modeler might want us to make.

[6] Here is as good a place as any to note that I will go back and forth in the text referring to the
people here who have the various psychological disorders (D), as both "people with D" and "D
people" (or sometimes just "Ds"). There is an ongoing lively debate, both within disability
studies and among those with various disorders, about the best language to use. "D people"
worryingly connotes to some that they are reduced to or defined by their disorder, whereas some
people with D *want* to be so defined, as that signals they are a member of a distinctive
community (and "people with D" connotes to them that they have a disease). For some
discussion of the relevant issues, see https://www.parents.com/health/special-needs-
now/should-we-say-with-autism-or-autistic-heres-why-it-matters/. I want to stay neutral on
this debate, so I will, as I say, go back and forth in such labeling (probably irritating both
sides in the process).

so that those students can most effectively learn in their distinctive ways. Identifying how students learn is, most fundamentally, a matter of *teacher empathy*, of teachers trying to understand from their students' own perspectives what it's like to be those students and learn in the way they do (Long et al. 2007; Cooper 2011). Consequently, constructing the teaching environment to enable individual learning may include: forming learning clubs and support groups with other students; providing model answers for exams; marking students' papers in front of them (so as to teach them via the grading); providing alternatives to students taking notes by dictation (Long et al. 2007); and reframing the sorts of questions teachers ask their students (switching from being on the lookout for deficits needing remediation to being more active and interested listeners). This switch helps enable students to construct "a *learning identity*...that emphasize[s] their intelligence, verbal skill, curiosity, and learning potential" (Dudley-Marling 2004: 488; emphasis added). The ultimate aim, via empathic discovery, is to individualize instruction in a way that enables these students to develop and flourish *as the reading-ready agents they already are*.

We now have a skeletal template for proceeding. In what follows I want to fill in some details of how it might apply to the accountability-"disabled." The key paradigm shift involves viewing psychologically disordered agents as just *differently abled* for accountability from those who are non-disordered. The aim, therefore, is not alteration but accommodation. Accomplishing it requires a much more expansive and intimate approach than in Glover's Project of Understanding. One has to come to see those with psychological disorders as what I will call *resonantly intelligible*.

To get there, start with what the teachers of dyslexics deploy: *empathy*. This requires more than mere cognitive understanding. The teachers of the dyslexic, after all, have to do more than just understand why their students "read funny," where in so doing they simply identify the causal explanation for their difficulty that traces back to, and stops at, an impairment. They need in addition to figure out how, *from the inside*, words and letters actually appear to their students, in order to design the best methods to engage with them as ready readers. So too when it comes to the accountability-"disabled," what has to be figured out is not merely what causes the behavior in question—whether it is their moral identity or their disorder, as Glover puts it—but instead the respects in which that behavior makes sense to them, the sense in which they view what they are doing as *intelligible*.

Those who view their actions as intelligible—as making sense given who they are and what they want—are precisely viewing those actions as things for which they are *accountable*, as behavior in fact guided by their moral identities and for which they view themselves as vulnerable at least to their *own* reactive attitudes (e.g. guilt and pride), depending on how well they execute those moral identities. There is plenty of empirical evidence for this point, the most overwhelming of which comes from studies on those with clinical depression. They tend to view themselves, even while in the throes of depression, as accountable for what they do, and so they feel guilty for what they deem to be *their own* failures of motivation (a feeling that tends to spiral them down even further into depression as a result; see Berrios et al. 1992; O'Connor et al. 2002; and Ghatavi et al. 2002). Other evidence about the sense of accountability and guilty feelings may be found with respect to those with eating disorders (e.g., Bybee et al. 1996) and those with Obsessive-Compulsive Disorder (e.g., Shafran, Watkins, and Charman 1996; Mancini and Gangemi 2004).

Many people with psychological disorders thus view themselves as perfectly capable of being in the accountability community *if only its other members would treat them as such*, which people could and would do were the socially created and enforced responses and practices of accountability simply rejiggered to accommodate those with psychological disorders. In order to accommodate exempted agents within the accountability community, therefore, the non-exempt must first come to see the exempt *as the exempt see themselves*, namely, as *accountable*, as the proper targets of the full range of reactive attitudes.[7]

For many people, though, this may be tough to do, stuck as they are in their own perspectives and moral identities. Those without psychological disorders of course see their own moral identities as intelligible guides of their *own* behavior, and so they think of themselves as eminently accountable (again, as nearly all humans do), but that's easy because those identities are their *own* (and so eminently reasonable). To enable the necessary perspectival transition, then, we have to get the non-disordered to see themselves in those with psychological disorders, to come to recognize how they themselves *could well have had* the moral identities that those with various disorders have, and so come to recognize that they themselves might well have behaved as those with disorders did had they had their moral identities.

We are almost where we need to be. But as Glen Pettigrove notes, to *fully* make sense of some agential event, we need to situate it in the context of

---

[7] Thanks to David Beglin for helpful feedback on this idea.

what came before it and what will come after it. We need to see it as part of the agent's developed and persisting story, her ongoing *narrative*. Where we start and end the story significantly affects our assessment of it (Pettigrove 2007: 173). If we leave out an agent's back story—where she came from, what she's gone through, what she'll go through—it's all too easy to see a hero as a villain, or, perhaps worse, as an alien. Thus the more of an agent's ongoing story we can include in empathizing with her, the greater sense we can make of her current moral identity and behavior in the way that she does (Pettigrove 2007: 173).

In sum, for those accountability-"disabled" agents who nevertheless construe themselves as intelligible and accountable, those without such "disabilities" must come to (a) empathize sufficiently with them to come to see how what they are doing makes sense from their own perspectives, given their moral identities (*intelligibility*), and (b) appreciate how they themselves might have come to have had the moral identities of those exempted from accountability, and so appreciate why they themselves might well have done the very same things under the guidance of that moral identity (*resonance*). As this project crucially aims at getting the non-exempt to see themselves in the exempt, call it the *Project of Identification*.[8]

As it stands, this is a crudely drawn picture, just a first stab at what the social model might suggest for accountability-"disabilities." In what follows, I hope to clarify and develop it by seeing whether and how it might plausibly be applied to three psychological disorders that currently count as accountability-"disabling."

I think the Project of Identification is most promising, first, with those on the autism spectrum. Autism is a social communication disorder generated by a family of impairments, including impairments in interpreting others ("theory of mind"), impairments in identifying emotions (alexithymia), impairments in imagination and empathy, and, finally, rigidity and repetitive movements. Given these various impairments, it might be (and is) thought that people should view autistics as incapable of meeting the demands for, say, tact, friendliness, or respectfulness, and so as exempt from accountability in the social domains in which these things are demanded.

Viewing those with autism through the lens of the social model yields a very different result. To start, one needs to pay attention to the burgeoning autobiographical autism literature. In an open letter to parents, for example,

---

[8] Thanks to Olivia Bailey for helpful feedback on this idea.

Jim Sinclair (1993) offers some suggestions to parents of autistic children (having been one himself):

> You try to relate as parent to child, using your own understanding of normal children, your own experiences and intuitions about relationships. And the child doesn't respond in any way you can recognize as being part of that system. That does not mean that the child is incapable of relating. It only means you're assuming a shared system, a shared understanding of signals and meanings, that the child in fact does not share ... It takes more work to communicate with someone whose native language isn't yours ....
> (Quoted in Glover 2014: 132)

Temple Grandin has famously talked about the fact that she thinks more in pictures than in words, and the meaning of terms for her depends heavily on past associations. Others with autism say they have developed rituals and repetitive movements as a defense against the rapid pace of the world around them: "The constant change of most things never seemed to give me any chance to prepare myself for them. Because of this I found pleasure and comfort in doing the same things over and over again" (Williams 1999: 45; quoted in Glover 2014: 133–4). Those with autism sometimes report that being touched feels like an assault, as if someone else has control over their body. To look directly at someone has been said to raise the fear that one's own identity will be given over to that person. Emotions may course through one's body but find no expression except via seemingly strange or inexplicable movements or sounds (Glover 2014: 134–6).

In applying the Project of Identification to autism, the first step of the allistic (non-autistic neurotypicals) is thus to start learning its language. They then need to come to see themselves in the autistic, to empathically identify with them. How so? It should be obvious, given the reports just noted, that many people with autism have moral identities that guide their behavior. They value peace, calm, respect for personal space, and the comforts of familiarity. But so do many people *without* autism! And so at least at the level of moral identities, the allistic's identification with the autistic should be as easy as it is with many non-autistics.

Of course, some of those with autism *express* their moral identities in non-standard ways. But so what? As long as one begins from the assumption that those with autism are abled, just *differently*, for accountability, then their "different" ways of executing their moral identities are irrelevant to their accountability-eligibility. All that matters for eligibility is the having and executing of a moral identity.

Notice, then, how the Project of Identification works. It moves the non-exempt to shift *their own* default exempting emotional stance toward those with autism—from non-reactive to reactive—and thus to accommodate the autistic into the accountability community as they are. They do so by coming to see that those with autism are *already* accountable agents, just in non-standard (but not "disabling") ways. This may mean treating autistic people as accountable in non-standard ways as well (just like teaching the dyslexic may require non-standard teaching methods), but exempting them is off the table. Exemption is thus a problem of the *non*-disordered, not the disordered, so the change that's required is a perspectival shift in the non-disordered. But this should be relatively easy for them to achieve when it comes to autism.

What, though, about schizophrenia? Throughout its history, the "kind" and concept of schizophrenia have undergone dramatic changes. It is actually just a motley set of symptoms, some of which have been emphasized over others at different times, and for different purposes (Boyle 1990). What Ian Hacking astutely notes is that, once those diagnosed with schizophrenia begin receiving medications and come to internalize the classification, they often start to view the illness as "other," or as "an evil agent," and so they come to attribute to the *illness* their "stupid, or gross, unfeeling, or simply crazy actions" (Hacking 1999: 113). Thus, schizophrenic patients sometimes offer the following sorts of descriptions of their auditory hallucinations and thought insertions: "Thoughts come into my head like 'Kill God'. It's just like my mind working, but it isn't. They come from this chap, Chris. They're his thoughts" (Frith 1992: 66; quoted in Campbell 1999: 609). More generally:

> [P]atients report that they feel the thoughts which occur in their heads as not actually their own. They are not experienced as thoughts communicated to them... but it is as if another's thoughts have been engendered or inserted in them. One of our patients reported physically feeling the alien thoughts as they entered his head and claimed that he could pin-point the point of entry!
>
> (Cahill and Frith 1996: 278; quoted in Campbell 1999: 610)

The Project of Identification has us working our way into an exempted person's perspective, within which she presumably already has a first-personal sense of her own accountability-status. Unfortunately, that sense is precisely what is missing in some people with schizophrenia. What we have to do first, then, is *enable* that first-personal sense of accountability, by getting the person acting on what she deems "alien" thoughts to come to see them as her own, and so view the actions those thoughts call for as manifesting her moral identity. We could thus subsequently treat her as accountable for those manifestations.

Following the Hacking diagnosis, her current misattribution of those thoughts is likely the result of her having internalized the diagnosis of schizophrenia, so perhaps this diagnosis needs to be removed or rejected. Of course, the internalization may persist regardless. In such a case, we should just alter the way we respond to her in normative domains, namely, by straightforwardly *treating* her as accountable for what she does under the directive of the "alien" thoughts, that is, by treating her actions as in fact attributable to her and so as manifesting her moral identity. This would involve directly switching our default emotional stance to her, from non-reactive to reactive. Suppose, then, that she were to attempt to act on an "inserted" directive to kill someone she was "told" was the devil, aiming to protect the rest of us from his evil. To get her to attribute that action to her "own" moral identity, we would have to treat it as *already* attributable to her by responding to her with gratitude and admiration for showing us an outstanding quality of will in saving our lives.

This approach strikes me as absurd. What's particularly worrisome about it is that there is a crucial *reason* that she simply can't seem to recognize for not killing this person; namely, that he isn't the devil! How can we treat someone as accountable who is incapable of recognizing such reasons?

Well, maybe we should remind ourselves that those in the accountability community often do exactly this! After all, everyone is irrational to some extent. We humans have well-known confirmation biases, anchoring biases, and make the fundamental attribution error, to name just a few. But these irrationalities don't disqualify us from accountable agency; indeed, they are just the types of things that are blamable. Why? Because *we should have known better*. The non-exempt might thus view our schizophrenic agent above as simply further along on that same spectrum of irrationality than most other people. Perhaps she should have known better than to believe what was directed by the "alien thought insertion." She might, then, be accommodated into the accountability community by our lowering the bar for what counts as culpable ignorance and *blaming* her for killing the person she took to be the devil.

If this is the way to accommodate those with schizophrenia into the accountability community, then it is less absurd than *immoral*. Glover agrees. In considering whether we should hold schizophrenic people responsible, he remarks that to "have these [responsibility] responses [to them] seems *unfair*, for all the reasons that make it doubtful that the behavior reflects the person rather than the illness" (Glover 2014: 376; emphasis mine).

While eliminating the "disability" of those with autism by viewing them through the lens of the social model seemed easy and fruitful, it is much more difficult with respect to schizophrenia, leading us down one path toward absurdity, and down the other path toward immorality. But as we will now see, when it comes to applying this approach to *psychopathy*, there is no plausible way to get the machinery off the ground.

Psychopathy of course comes in degrees. But those who have written about it in the responsibility literature focus essentially on those at the far end of the spectrum, those truly without empathy, those who take advantage of, lie to, and hurt others without compunction. I will do the same. Given these extreme psychopaths' fundamental empathic impairments, the majority of theorists view them as exempt from accountability. These psychopaths can't take up the perspectives of others to see and feel what things are like for them, so they fail to take seriously others' ends and interests as valuable, as mattering, as emotionally infused, and so as worth protecting and respecting. And this is, I believe, because they cannot see their *own* pursuits as valuable, mattering, or emotionally infused (see Cleckley 1982; Watson 2013). Psychopaths are wildly imprudent, often doing things that will completely set back their own interests, and not knowing or caring one whit about it. They lack empathy for *themselves*, and this makes them both morally *and* prudentially impaired (Shoemaker 2015: Chs. 5–6).

The Project of Identification is a rich empathic project, and as such it won't work with extreme psychopaths because we are being asked to robustly empathically identify *with the unempathic*. That is, we are being asked to feel what it's like for people who cannot feel what it's like for *anyone* (including themselves). This is why psychopaths remain at least partially alien to non-psychopaths.

Now you may think you can nevertheless work your way into identifying with psychopaths, perhaps step-by-step.[9] For instance, psychopaths do have desires and feel pains and pleasures, so you do share that with them. But can

---

[9] Thanks to Hanna Pickard for pushing me in this direction.

you imagine being someone who is engaged in the single-minded pursuit of whatever desire crops up? Well, perhaps you can get that far too, as you were a teenager once. But now imagine that nothing really matters to you, nothing is *worth* pursuing. Then try imagining someone pleading with you to stop doing what you're doing, screaming "That hurts!", yet being so cold to that plea that it doesn't register with you either emotionally (sympathetically) or as a putative reason to refrain, that their pain seems just another boring fact akin to how wide some smudge is on your car's windshield or how many ants exist in New Mexico. Next, try imagining seeing that person's pain as a reason to *cause* it, seeing reasons of amusement in horrific acts, but seeing no reasons against cruelty. Try imagining further having no emotional engagement with "friends" and family, having no one you'd sacrifice a thing for, not feeling any real aversion to fearsome threats, and more. It grows harder and harder, I suspect, to achieve anything like genuine or robust empathy with the psychopathic. And so the gulf remains. I cannot see any way of bridging it.[10]

The problem is that extreme psychopaths seem to lack a moral identity altogether. They don't or can't really care about—have any emotional investment in—anything, let alone other people. But perhaps that's all irrelevant. Perhaps it doesn't matter that we can't work our way into the heads of extreme psychopaths. Perhaps we should just treat them as accountable regardless, reacting to their cruelty with resentment and indignation.[11] Unfortunately, doing so raises exactly the same moral problem we had in doing this with some schizophrenic killers: How could it be fair to blame psychopaths when what they do seems to manifest their disorder, not their identity?

*   *   *

Trying to eliminate the moral costs of excuses-exemptions by viewing those treated with accountability-"disabilities" through the lens of the social model, while very promising for autism, faces genuine, perhaps insurmountable, difficulties when applied to other disorders. And even were its application successful, we would have brand new moral problems on our hands. Damned if we do; damned if we don't.

---

[10] There are now laboratory aids to help generate a temporary glimpse into what it's like to be a psychopath, but the foreignness is only made more dramatic, I think, and non-psychopaths can't get there on their own. See https://www.chronicle.com/article/The-Psychopath-Makeover/135160 (Thanks to Monique Wonderly for the pointer.)

[11] For those sympathetic to this move, see Talbert 2008 and 2012; Scanlon 1998; and Harman 2011 and 2019.

2.5 Interpersonal Relationships and the
Accountability Community

People with various psychological disorders who are excused-exempted—
viewed with a default emotionally unengaged stance—are deprived of
important moral goods, discriminated against on the basis of their disorders.
Seeing them through lens of the medical model of disability (via the Project
of Understanding) has the non-disordered changing their default emotional
stances only after those with disorders have been successfully treated and
"fixed," that is, *made* accountable. Seeing them through the lens of the social
model of disability (via the Project of Identification) has the non-disordered
changing their default emotional stances to disordered people as a function
of their newfound identification with the disordered. As the non-disordered
can suddenly *see* what it's like to have the relevant disorder, and they can see
how *they themselves* might well have been disordered, their emotional
engagement—a sign of shared membership in the accountability
community—ought directly to flow. Unfortunately, the medical model of
disability preserves discrimination against untreated agents, or agents who
can't be "fixed," and it also fails to recognize or take seriously the many
environmental and political constituents of "disability." And whereas the
social model of disability avoids both problems, its normative recommendation to accommodate the excused-exempted as they are into the accountability community leads to absurdity (gratitude for killing "the devil") or
new immorality (e.g., unfairly treating people as accountable for things that
just aren't properly attributable to them).

Rock or hard place? The key to avoiding both is to distinguish between
the interpersonal and accountability communities. That is to say, we need to
abandon the widely shared core Strawsonian assumption that got us started.
These are just different communities with different membership conditions
(cf., Kennett 2009; Wallace 1994 [Ch. 2] and 2014). Interpersonal life is
indeed shot through with exchanges of Strawson's participant attitudes. But
most of the participant interpersonal attitudes are actually not the reactive
attitudes responding to accountable agency.[12] While the latter include
resentment, indignation, gratitude, and guilt, the former include shared

---

[12] Note also Kennett (2009: 12): "While one might need to qualify as an autonomous agent to
be a fit subject for the reactive attitude of resentment, it is not at all obvious that full autonomy is
needed to fit one for the many other participant reactive attitudes, or must be present to ground
the moral demand for respect and goodwill."

affective experiences of friendliness, rooting, love, amusement, enjoyment, glee, joy, grief, and sadness. The participant attitudes are simply about emotional engagement, communion, and connection with other people. They are the glue of humanity.[13]

When it comes to enabling the conditions of accountable agency, though, susceptibility to such participant attitudes alone is just insufficient, and may in fact be irrelevant. Jeanette Kennett puts it well: "[W]hat is valuable in our relations with each other, and our moral standing within those relations, is not given wholly by the features that make us accountable agents" (Kennett 2009: 12). We do not make people accountable agents merely by opening ourselves up emotionally to being their friends, enjoying their company, loving them, rooting for their success, or being amused together. And those incapable of emotionally engaged interpersonal life (*Star Trek*'s Data?) may nevertheless be accountable agents. While the circles of the two communities intersect, neither is anything like a subset of the other.

What, then, are the conditions for membership in each community? Both communities have membership by degrees, as there are numerous features and capacities involved in both. I will nevertheless speak here in more stark on-off terms, simply to get the main ideas across. To be a member of the interpersonal participant community, one must be *resonantly intelligible to others*, establishable in principle by others via the Project of Identification. To be a member of the accountability community, alternatively, one must have certain individual agential capacities, for example, *to have and be able to execute a sufficiently deep moral identity*, something discoverable in principle by others via the Project of Understanding.[14]

---

[13] Are they extendable to nonhuman animals? A few are, perhaps, in proto-form, e.g., a kind of shared joy with one's dog, a kind of shared affection between a biologist and her learned chimpanzee. Because there are so many participant attitudes, they establish a wide spectrum of emotional engagement. But the shared emotional engagement I have in mind is mostly going to obtain just human-to-human. This is because, as I will argue, emotional engagement is really a function of identifiability, and with very few exceptions we humans can identify only with other humans, precisely because it involves tapping into moral identities and values, which are essentially exclusive to human beings. This also means that the *kind* of shared emotional engagement I have in mind is also (probably, mostly) exclusive to humans. Chimpanzees certainly romp and play with each other, and they *seem* to display a kind of shared amusement in so doing. But is it actually amusement? Again, proto-amusement, perhaps, but I just don't know what amusement is like *for* chimpanzees, so I can't say whether it's of our kind. I'm inclined to the Wittgensteinian thought here: "If a lion could speak, we could not understand him."

[14] Inside baseball footnote: This may sound like I think that accountability is response-independent, which would be contrary to what I've argued elsewhere (e.g., Shoemaker 2017), but that would be a mistaken reading. My view is that the objective properties (e.g., capacities)

The story of interpersonality is most in need of explanation, I recognize. While accountability is more a familiar matter of call and response, a kind of conversation between doers and reactors (see, e.g., Macnamara 2013 and 2015; McKenna 2012), the participant stance is more about shared emotional vulnerability. This means that people may be accountable agents, making a conversational (calling-out) gambit, without anyone hearing or responding, whereas the participant interpersonal stance requires actively *shared* vulnerabilities. You may treat me as an object to be "managed or handled," and that may well hurt me emotionally, but your objective stance toward me is incompatible with our standing in the participant community together. To do that requires that you at least have the *ability* to engage emotionally with me. This means that the conditions of membership in the interpersonal community are themselves a partial function of the capacities of *identifiers*, people who are capable of successfully deploying the Project of Identification by empathizing with potential members and seeing them as resonantly intelligible.[15]

The history of humanity and morality is in part a history of the outward expansion of emotional identification.[16] This is a matter of degree: the more that people are resonantly intelligible to one another, the more relatable and less alien they are to one another. The Project of Identification has primarily been successful just with other humans.[17] Of course, I have suggested above that we may have genuine difficulty identifying with psychopaths. But I think we are capable of seeing virtually all other humans as resonantly intelligible, even those with profound intellectual disabilities.[18] There is surely a certain kind of *status*—a participant, interpersonal status—attached to being empathizable in this way, one whose edges extend to nearly all of humanity.

So now here is what to say about accountability: Among the things I might discover in identifying with you and then returning to my own

that make someone accountable are themselves response-dependent, counting as accountability-making properties in virtue of being the object of fitting responses by those with refined sensibilities. That's all compatible with what I say in the text.

[15] This point is clearly about identifiers' *capacities*. Whether or not they actually exercise them and *do* empathize and identify with others, well, that implicates a different long, familiar, and tragic story. See Gaita (2002).

[16] See, e.g., the very different work by Singer (1981), and Buchanan and Powell (2018).

[17] It *may* also be limited to other humans, although I won't take a definitive stance on that point here. See fn. 13 above.

[18] See the groundbreaking work by Eva Feder Kittay (2005, 2019); on her emotional engagement with her own profoundly intellectually disabled daughter Sesha. It is just different in kind, it seems to me, than the sort of emotional interchange one can have with nonhuman animals, and I share Kittay's sense that it's insulting to think otherwise, although I realize how contentious that view is.

perspective is that your moral identity is actually rather fragile, shallow, corrupted, or broken, and so you may be mistaken in thinking that what you do is attributable to you. I can, in other words, see myself in you and so see why from your perspective you may *think* you are an accountable agent, but in a way that subsequently allows me to understand from my *own* perspective (having seen your story in wide narrative scope) why you *aren't* in fact accountable, or why you are less accountable than you think. In such cases, I may well extend the moral *benefit* of exemption to you.

What I have said may seem odd or unclear, but the basic idea is actually illustrated beautifully in Gary Watson's (2004: 219–59) influential discussion of Robert Alton Harris. Harris murdered two teenagers in cold blood, then calmly ate the fast food they had ordered, joking that he ought to dress up like a police officer to inform their parents. Once he was on death row, his fellow prisoners hated him, calling him a complete "scumbag," someone at the bottom of the human barrel. Now when we assess his murderous actions at this point, he looks like the perfect candidate for our strongest condemnation, seeming to have an evil moral identity that he clearly manifested in action. Watson's worry, though, is that, on the core Strawsonian assumption, his extreme evil renders him *alien* to us, both excluding him from the participant interpersonal community and exempting him from accountability, the latter of which is absurd.

But Watson then has us engage in the Project of Identification (albeit not under that name), charitably expanding the narrative scope of Harris's life. It turns out he was the product of an intensely abusive childhood, repeatedly beaten and sexually assaulted, one of the most horrific upbringings one can imagine. Once we hear this fuller story, Watson suggests, our overall response toward Harris becomes ambivalent. Our blame doesn't disappear, exactly, but it now sits uneasily with Harris's full biography, which "forces us to see him as a *victim*," and evokes "conflicting responses" (Watson 2004: 244; emphasis in original). "The sympathy toward the boy he was is at odds with outrage toward the man he is" (Watson 2004: 244). What our empathic identification with him does is allow us to see Harris's cruelty as in fact "an *intelligible* response to his circumstances," which "gives a foothold not only for sympathy, but for the thought that if *I* had been subjected to such circumstances, I might well have become as vile. What is unsettling is the thought that one's moral self is such a fragile thing" (Watson 2004: 245; first emphasis added; second emphasis in original). Understanding the *fragility* of his moral identity, once we return to our own perspective, leads us to see Harris as being less of an accountable agent than we had originally thought, while identifying with him through the broader lens of his awful childhood also brings us closer to him as a fellow human. We see him as less alien now, and more as a broken version of one of *us*.

On my view, the "ambivalence" of our response to Harris raises no tensions. Our engaged sympathy for him is a participant attitude; our now-dampened outrage is an accountability attitude. It is possible both to see someone as a fellow human, someone with whom we can emotionally engage with on many fronts, and to see him as having a broken moral identity that leaves him a less-than-fully-accountable agent.

Here is how to resolve the moral costs motivating this paper: Recognize that many of the goods withheld from those exempted from accountability are actually goods available within *the participant stance alone*, regardless of accountability status. They are a function of the extent to which *we are able to identify* with our fellows, not a function of our fellows' individual accountability capacities. The persistent mistake has been to assume that various psychological disorders *both* excuse people from interpersonal emotional life *and* exempt them from accountability, whereas only the latter may be true. Consequently, eliminating discrimination against them is entirely a matter of a undergoing a paradigm shift in how they are viewed, a function of seeing them through the lens of the social model of disability by actually carrying out the Project of Identification in order to establish their accommodation and inclusion in interpersonal life.

Let me illustrate by returning to the two specific moral costs of exemption raised at the beginning. When I identify with you, that is, when I see you as resonantly intelligible, then we have become vulnerable to emotional engagement with one another. This is all that's necessary for us to enjoy each other's company, root for each other's successes, be amused or heartbroken alongside one another, be friends, fall in love, and so forth. And this is true even if I discover, among other things, that your moral identity is shallow, corrupt, or broken, or that your actions manifest your illness, not your moral identity, and so you are not an accountable agent. If we separate out the conditions for interpersonality from the conditions for accountability, we can exempt people from accountability without emotionally starving them.[19]

The Project of Identification can also ameliorate worries about the lack of regard and recognition. What I am doing in robustly identifying with you is

---

[19] In a way, I am extending Hanna Pickard's (2011) "responsibility without blame" one step further, to "interpersonality without blame *or responsibility*." Thanks to Joshua May for this suggestion.

in fact *acknowledging* you. When I empathically identify with you, and I appreciate how I might have wound up with a moral identity very much like yours, what I recognize is something like our *common human nature* (cf. Gaita 2002): We have both arisen out of and share the same basic human building materials that may be shaped, developed, twisted, and broken in familiar ways. We are *equals* in this sense. Acknowledging this status in others is what I have elsewhere called *pure regard* (Shoemaker 2015: Ch. 3). It involves taking you seriously as a fellow, recognizing our fundamental moral equality, and thus perceiving that your interests provide at least putative reasons for me to take into account from within my own deliberative framework. But this is also just the sort of recognition Glover argues is essential, recall, for the *development* of moral agency, something that enables people to become more securely aware of who they are and to create their own deep moral identities (Glover 2014: 309). It is the sort of recognition that was crucially missing in the formative years of many of those with ASPD that he interviewed. This type of recognition—acknowledgment—is a good that can indeed be distributed to those within the interpersonal participant community alone, even if they are exempted from accountable agency.[20]

By distinguishing between the conditions of membership in the interpersonal and accountability communities, we can ameliorate a significant amount of the actual moral costs there would be by continuing to view their conditions as identical. The key point here is that exemption isn't necessarily "disabling"; that is, agents can be exempt from accountability without being discriminated against thereby, for as long as they are emotionally relatable from the participant stance (i.e., they are resonantly intelligible), they can have equal access to the significant goods inherent to the interpersonal community. Further, by continuing to exempt some people from accountability who are nevertheless resonantly intelligible, we can avoid the alternative moral costs of holding people accountable whose actions are attributable to their disorders and not to their moral identities.

---

[20] There is of course a deeper type of recognition unavailable to those exempted from accountability, the type of Darwallian "second-personal status" of only those capable of recognizing the relevant moral equality and legitimacy of accountability demands persons can make of one another (Darwall 2006). There are likely distinctive goods attached to this status that are indeed being withheld from people exempted from accountability. But, first, these are not the goods of acknowledgment Glover thinks so important to moral development and human engagement, and, second, withholding something from someone incapable of enjoying it as a good may not be a deprivation in the first place. Thanks to Shaun Nichols for helping me think through this point.

2.6 Conclusion

What I have tried to show is that the "disabilities" of those who are excluded-exempted comes from the *exclusion* side of the hyphen, that is, from their being excluded from interpersonal emotional life. Being exempted from accountability, all on its own, isn't "disabling." To ameliorate the relevant "disabilities," then, those who have been excluded from interpersonal life for various psychological disorders need to be viewed through the lens of the social model of disability and accommodated as they are into that community, via a change in their fellows' perspectives and emotional openness. This will enable their equal access to goods such as fellow-feeling and acknowledgment.

Nevertheless, "medicalizing" may still be perfectly appropriate for those impaired for accountability. The Project of Identification, after all, may reveal to us once we return to our own perspective that someone's moral identity—which we ourselves might have come to have had if things had gone differently—is still insufficiently deep to ground accountable agency. Here it still seems appropriate to treat and fix these broken or wounded agents as best we can, especially given that we can do so without depriving them of the significantly valuable goods of interpersonal human life.[21]

References

Barnes, Elizabeth. 2016. *The Minority Body*. Oxford: Oxford University Press.

Bennett, Jonathan. 1980. "Accountability." In Zak van Straaten, ed., *Philosophical Subjects* (Oxford: Clarendon Press): 14–47.

[21] For helpful questions and comments on my presentations of the ideas in this chapter, I'm grateful to audiences at: the UAB Workshop on Responsibility and Mental Disorder (including especially Chandra Sripada, Walter Sinnott-Armstrong, Holly Kantin, and Natalia Washington); the Princeton University Center for Human Values (including especially Monique Wonderly, Michael Smith, Shaun Nichols, Mark Van Roojen, Elizabeth Harman, Hanna Pickard, Mitch Berman, Sam Preston, Stephanie Beardman, Jason White, and Lori Gruen); the Conference on "What Can Philosophy Learn from Disability?" at Monash University (including Jeanette Kennett, Eva Feder Kittay, Stephanie Elsen, Michael Smith again, and Stephanie Collins), and at a Cornell University department colloquium. I'm also grateful to several people who read and commented on earlier drafts of the paper, including Daniel Putnam, Monique Wonderly, Elizabeth Barnes, August Gorman, Drew Schroeder, Samuel Lundquist, Anneli Jefferson, David Beglin, and Olivia Bailey. Finally, thanks to Matt King and Joshua May for penetrating and extremely helpful comments on a penultimate draft.

Berrios, G.E., et al. 1992. "Feelings of Guilt in Major Depression: Conceptual and Psychometric Aspects." *The British Journal of Psychiatry* 160: 781–7.

Bhugra, Dinesh. 1989. "Attitudes towards Mental Illness: A Review of the Literature." *Acta Psychiatrica Scandinavica* 80: 1–12.

Boyle, Mary. 1990. *Schizophrenia: A Scientific Delusion?* London: Routledge. (2nd edition published in 2014.)

Buchanan, Alan, and Powell, Russell. 2018. *The Evolution of Moral Progress*. Oxford: Oxford University Press.

Bybee, Jane, et al. 1996. "Guilt, Guilt-Invoking Events, Depression, and Eating Disorders." *Current Psychology* 15: 113–27.

Cahill, Connie, and Frith, Christopher. 1996. "False Perceptions or False Beliefs? Hallucinations and Delusions in Schizophrenia." In P.W. Halligan and J.C. Marshall, eds., *Method in Madness: Case Studies in Cognitive Neuropsychiatry* (Hove: Psychology Press), pp. 267–91.

Campbell, John. 1999. "Schizophrenia, the Space of Reasons, and Thinking as a Motor Process." *The Monist* 82: 609–25.

Cleckley, Hervey. 1982. *The Mask of Sanity* (Rev. Ed.). St. Louis: Mosby.

Cooper, Bridget. 2011. *Empathy in Education: Engagement, Values and Achievement*. London: Bloomsbury Publishing.

Darwall, Stephen. 2006. *The Second-Person Standpoint*. Cambridge, MA: Harvard University Press.

Dudley-Marling, Curt. 2004. "The Social Construction of Learning Disabilities." *Journal of Learning Disabilities* 37: 482–9.

Elliott, Julian G., and Gibbs, Simon. 2008. "Does Dyslexia Exist?" *Journal of Philosophy of Education* 42: 475–91.

Fischer, John Martin and Ravizza, M. 1998. *Responsibility and Control: A Theory of Moral Responsibility*. Cambridge: Cambridge University Press.

Fricker, Miranda. 2016. "What's the Point of Blame? A Paradigm-Based Explanation." *Nous* 50: 165–83.

Frith, Christopher D. 1992. *The Cognitive Neuropsychology of Schizophrenia*. Hillsdale, NJ: Erlbaum.

Gaita, Raimond. 2002. *A Common Humanity*. London: Routledge.

Ghatavi, Kayhan, et al. 2002. "Defining Guilt in Depression: A Comparison of Subjects with Major Depression, Chronic Medical Illness and Healthy Controls." *Journal of Affective Disorders* 68: 307–15.

Glover, Jonathan. 2014. *Alien Landscapes?* Cambridge, MA: The Belknap Press of Harvard University Press.

Hacking, Ian. 1999. *The Social Construction of What?* Cambridge, MA: Harvard University Press.

Harman, Elizabeth. 2011. "Does Moral Ignorance Exculpate?" *Ratio* 24: 443–68.

Harman, Elizabeth. 2019. "Moral Testimony Goes Only So Far." *Oxford Studies in Agency and Responsibility* 6: 165-185.

Hayes, Jeanne, Boylstein, Craig, and Zimmerman, Mary K. 2009. "Living and Loving with Dementia: Negotiating Spousal and Caregiver Identity through Narrative. *Journal of Aging Studies* 23: 48–59.

Kennett, Jeanette. 2009. "Mental Disorder, Moral Agency, and the Self." In Bonnie Steinbock, ed., *The Oxford Handbook of Bioethics*. Oxford: Oxford University Press.

Kittay, Eva Feder. 2005. "At the Margins of Moral Personhood." *Ethics* 116: 100–31.

Kittay, Eva Feder. 2019. "Love's Labor: Essays on Women, Equality, and Dependency." London: Routledge.

Long, Louise, MacBlain, Sean, and MacBlain, Martin. 2007. "Supporting Students with Dyslexia at the Secondary Level: An Emotional Model of Literacy." *Journal of Adolescent & Adult Literacy* 51: 124–34.

Macnamara, Coleen. 2013. "Taking Demands Out of Blame." In D. Justin Coates and Neal Tognazzini, eds., *Blame: Its Nature and Norms* (New York: Oxford University Press), pp. 141–61.

Macnamara, Coleen. 2015. "Reactive Attitudes as Communicative Entities." *Philosophy & Phenomenological Research* 90: 546–69.

Mancini, Francesco, and Gangemi, Amelia. 2004. "Fear of Guilt from Behaving Irresponsibly in Obsessive-Compulsive Disorder." *Journal of Behavior Therapy and Experimental Psychiatry* 35: 109–20.

Matsakis, Aphrodite T. 2014. *Loving Someone with PTSD: A Practical Guide to Understanding and Connecting with your Partner after Trauma*. Oakland, CA: New Harbinger Publications.

McGeer, Victoria. 2019. "Scaffolding Agency: A Proleptic Account of the Reactive Attitudes." *European Journal of Philosophy* 27: 301–23.

McKenna, Michael. 2012. *Conversation and Responsibility*. New York: Oxford University Press.

Nabors, Nina, Seacat, Jason, and Rosenthal, Mitchell. 2002. "Predictors of Caregiver Burden Following Traumatic Brain Injury." *Brain Injury* 16: 10-39-1050.

O'Connor, Lynn E. et al. 2002. "Guilt, Fear, Submission, and Empathy in Depression." *Journal of Affective Disorders* 71: 19–27.

Oliver, Mike. 1996. *Understanding Disability: From Theory to Practice.* London and Basingstoke: Palgrave Macmillan.

Oliver, Mike, and Barnes, Colin. 2012. *The New Politics of Disablement.* London and Basingstoke: Palgrave Macmillan.

Pettigrove, Glen. 2007. "Understanding, Excusing, Forgiving." *Philosophy and Phenomenological Research* 74: 156–75.

Pickard, Hanna. 2011. "Responsibility without Blame: Empathy and the Effective Treatment of Personality Disorder." *Philosophy, Psychiatry, and Psychology* 18: 209–23.

Rabkin, Judith. 1974. "Public Attitudes toward Mental Illness: A Review of the Literature." *Schozophrenia Bulletin* 10: 9–33.

Russell, Paul. 2013. "Responsibility, Naturalism, and 'the Morality System'." *Oxford Studies in Agency and Responsibility* 1: 181–204.

Scanlon, T.M. 1998. *What We Owe to Each Other.* Cambridge, MA: The Belknap Press of Harvard University Press.

Scanlon, T.M. 2008. *Moral Dimensions.* Cambridge, MA: The Belknap Press of Harvard University Press.

Shafran, Roz, Watkins, Elizabeth, and Charman, Tony. 1996. "Guilt in Obsessive-Compulsive Disorder." *Journal of Anxiety Disorders* 10: 509–16.

Shakespeare, Tom. 2013. "The Social Model of Disability." In Lennard J. Davis, ed., *The Disability Studies Reader*, 4th ed. (London: Routledge), pp. 214–21.

Shoemaker, David. 2007. "Moral Address, Moral Responsibility, and the Boundaries of the Moral Community." *Ethics* 118: 70–108.

Shoemaker, David. 2015. *Responsibility from the Margins.* Oxford: Oxford University Press.

Shoemaker, David. 2017. "Response-Dependent Responsibility; Or, a Funny Thing Happened on the Way to Blame." *Philosophical Review* 126: 481–527.

Sinclair, Jim. 1993. "Don't Mourn for Us." *Our Voice*, newsletter of the Autism Network International. Vol. 1, no. 3.

Singer, Peter. 1981. *The Expanding Circle.* Oxford: Clarendon Press.

Sneddon, Andrew. 2005. "Moral Responsibility: The Difference of Strawson, and the Difference It Should Make." *Ethical Theory and Moral Practice* 8: 239–64.

Sommers, Tamler. 2012. *Relative Justice.* Princeton, NJ: Princeton University Press.

Stern, Lawrence. 1974. "Freedom, Blame, and Moral Community." *Journal of Philosophy* 71: 72–84.

Strawson, P.F. 1962. "Freedom and Resentment." Reprinted in Gary Watson, ed., *Free Will*, 2nd ed. (Oxford: Oxford University Press, 2003), pp. 72–93.

Talbert, Matthew. 2008. "Blame and Responsiveness to Moral Reasons: Are Psychopaths Blameworthy?" *Pacific Philosophical Quarterly* 89: 516–35.

Talbert, Matthew. 2012. "Moral Competence, Moral Blame, and Protest." *The Journal of Ethics* 16: 89–109.

Vargas, Manuel. 2013. *Building Better Beings*. Oxford: Oxford University Press.

Wallace, R. Jay. 1994. *Responsibility and the Moral Sentiments*. Cambridge, MA: Harvard University Press.

Wallace, R. Jay. 2014. "Emotions and Relationships: On a Theme from Strawson." *Oxford Studies in Agency and Responsibility* 2: 119–42.

Watson, Gary. 2004. *Agency and Answerability*. Oxford: Oxford University Press.

Watson, Gary. 2013. "Psychopathic Agency and Prudential Deficits." *Proceedings of the Aristotelian Society* 113: 269–92.

Watson, Gary. 2014. "Peter Strawson on Responsibility and Sociality." *Oxford Studies in Agency and Responsibility* 2: 15–32.

Williams, Donna. 1999. *Nobody Nowhere: The Remarkable Autobiography of an Autistic Girl*. London: Jessica Kingsley.

Wolff, Jonathan. 2011. *Ethics and Public Policy*. London: Routledge.

3

Brain Pathology and Moral Responsibility

Anneli Jefferson

Cases of mental illness and neurosurgeons who alter people's brains in ways that make them do unspeakable things are common in the philosophical literature on moral responsibility, often used to test the limits of responsibility. A common thought is that, if one's action is the result of something messing with one's brain, be it a brain disease or a nefarious neurosurgeon, this counts at least as a *prima facie* excuse or exemption from being blameworthy (or praiseworthy) for that action. A thought experiment employed by Lawrie Reznek (1997) and discussed by Neil Levy (2007b) illustrates the common intuition that brain disorders excuse from moral responsibility: a boy, Billy, is born with a slow-growing brain tumor and becomes aggressive and criminal as the result of the tumor. Levy claims that on finding out about Billy's tumor, we would not blame him for his immoral behavior. Here is the case:

> As Billy grows, he develops a character which is formed, in part, by the presence of the tumor; it causes him to be aggressive and selfish. In his teens, he is involved in a string of increasingly serious crimes, culminating in a bungled bank robbery, hostage taking, and shoot out with police. Billy is fatally wounded. Now, it is clear that when his tumor is discovered at autopsy, we would cease blaming Billy for his vicious behavior.
>
> (Levy 2007b, 134)

My question in this chapter is whether we should cease blaming people like Billy for vicious behavior. In what way, if any, do brain pathologies such as a tumors excuse immoral behavior? Time and again, brain pathology has been taken to undermine responsibility-relevant characteristics such as control, whether in the context of addiction (as in Harry Frankfurt's (1971) example

Anneli Jefferson, *Brain Pathology and Moral Responsibility* In: *Agency in Mental Disorder: Philosophical Dimensions.*
Edited by: Matt King & Joshua May, Oxford University Press. © Anneli Jeffersonr 2022.
DOI: 10.1093/oso/9780198868811.003.0004

of the addict whose desire for the drug is effective because he is physiologically addicted) or brain tumors which undermine people's self-control (Burns and Swerdlow 2003). I will argue that the inference from brain pathology to reduced or absent responsibility is very indirect. Brain pathology matters to moral responsibility only in so far as it underlies and provides further evidence for psychological dysfunction that is relevant to responsibility.

However, according to a common intuition, brain tumors also matter because of the way they *cause* problematic dispositions and behavior. I argue that generally, causal history does not matter for moral responsibility; rather, responsibility rests on psychological capacities, regardless of how these are caused. However, the way psychological capacities change in brain disorders does have implications for moral responsibility, but in a way that has not been appreciated to date. Both in the case of classic brain disorders, such as tumors, and in the case of psychiatric illness, what matters to moral responsibility is not just the fact that a person's intrinsic psychological capacities change. Rather, responsible agency can also be affected by the way these intrinsic changes affect relational aspects of moral responsibility: it affects how successfully an individual's moral agency can be supported by their social environment. The resources we employ to find out how we should behave and to control our behavior are partly external; we rely on others to support our moral agency in numerous ways. When someone's moral psychology unexpectedly and sometimes drastically changes, both the individual and their social environment are unequipped to deal with these changes. This means that one's brain dysfunction can affect responsibility even if one's psychological capacities would still be sufficient for moral responsibility in the right kind of environment.

I provide a definition of brain pathology as well as a rough outline of the way mental illness (including brain pathology) can excuse in Section 3.1. In Section 3.2, I argue that the role of brain pathology for responsibility ascriptions should be an evidential one, knowledge of brain pathology can provide corroborating evidence for psychological dysfunction relevant to responsibility. I illustrate this with a case study. Finally, in Section 3.3, I consider the claim that brain disorders matter for moral responsibility because they change an individual's moral psychology in a way that is beyond their control. While control over psychological changes is not an excusing factor, as I show, brain disorders may mitigate moral responsibility because they confront individuals with new psychological deficits or urges for which their previous moral education and existing external and internal moral resources have not prepared them.

3.1 Preliminaries

3.1.1 Why Mental Disorders or Brain Disorders Might Excuse

The idea that mental or brain disorders can provide a reason to excuse or exempt individuals from responsibility is commonplace in the moral responsibility literature, though the exact reasons why this should be the case are frequently unclear. While the aim of this chapter is not to defend a specific theory of responsibility, it is necessary to make some minimal claims about what it takes to be morally responsible and why mental illness or brain disorder might impair one's moral responsibility. Rightly or wrongly, many philosophers have claimed that mental illness or brain disease undermine moral responsibility. This may happen in two key ways: when the ability to understand the nature and moral implications of one's actions is affected (and thus the ability to meet the knowledge or *epistemic condition* for moral responsibility), or when control of behavior is impaired, and one no longer meets the *control condition* for moral responsibility (cf. e.g. Brink and Nelkin 2013). Both conditions can be mapped onto a reasons-responsiveness account of moral responsibility, according to which an agent is responsible if they are responsive to moral reasons, where this involves both the ability to recognize moral reasons and to act on them (Fischer and Ravizza 1998).

The ability to recognize reasons is affected when psychiatric illness prevents individuals from understanding the permissibility of what they are doing because they are mistaken about external reality. For example, in the grip of psychosis patients might mistake non-threatening behavior as threatening, believe humans are aliens or robots and therefore sincerely and non-culpably take themselves to be acting in self-defense.[1] Knowledge can be affected more narrowly if illness prevents someone from understanding the wrongness of their actions. Some authors have argued this is the case for psychopaths (Fine and Kennett 2004; Shoemaker 2011).

Psychiatric illness might also affect individuals' control over action, by making impulse control significantly more difficult, thereby diminishing moral responsibility without completely removing it. On some accounts,

---

[1] Though see Broome, Bortolotti, and Mameli (2010) for some further complexities in these kinds of cases.

this is what happens in addiction.[2] Finally, mental illness can even lead to behavior that cannot be described as reasons-guided action at all, because it involves a reflex, sleepwalking, a compulsion, or similar cases in which control is undermined.

The question how frequently mental illness undermines individuals' capacity to recognize and respond to moral reasons for action is increasingly debated (King and May 2018; Arpaly 2005; Pickard 2015; Broome, Bortolotti, and Mameli 2010). For now, we can content ourselves with noting that in as far as one, or both, of the conditions of understanding and controlling one's actions are undermined by mental illness, moral responsibility is diminished. I have defined the capacity necessary for moral responsibility as reasons-responsiveness, which breaks down into a control condition and an epistemic condition. One other prominent account of moral responsibility defines responsible action in terms of deep, or real, selves (Frankfurt 1971): actions and desires have to be endorsed by an agent and reflect who they *really* are in order for the agent to be responsible for them. I will leave these positions to one side, as I think that—irrespective of whether one endorses these views—reasons responsiveness constitutes a necessary condition for responsible agency. I am also skeptical about the distinction between deep and non-authentic selves. Especially in the case of irreversible mental health conditions, it is morally problematic to claim that the disordered self is not the agent, as it forces us to discount the only self that the agent currently has (or is).

3.1.2 The Role of the Brain and the Nature of Brain Dysfunction

In order to assess the relevance of brain pathology to moral responsibility, we need to clarify the notion of brain pathology and its relation to mental disorders.

Paradigm cases of brain pathology are neurodegenerative diseases, brain trauma, and brain tumors, where there is a clearly identifiable physiological problem with the brain. But many psychiatrists and neuroscientists also assume that most, if not all, mental disorders involve brain pathology and see identifying these putative pathologies as an important task of the brain

---

[2] See Pickard (2015) for a convincing argument that in addiction, control is compromised but not undermined to the extent that responsibility is lacking.

sciences (Insel 2013; Cuthbert 2014). By contrast, others have argued for the claim that mental disorders and brain disorders are distinct categories and that a condition cannot be both a mental disorder and a brain disorder (Graham 2013). I hold the view that brain differences should count as dysfunctions if they reliably realize a psychological dysfunction. It is a further empirical question whether we will be able to find these kinds of brain anomalies for many mental disorders (Jefferson 2020); for example, people doubt that this will be possible for a condition like depression, which can take many different forms.

Whether one endorses an inclusive concept of brain disorder or not will depend on a number of issues in the philosophy of science which need not concern us here. The important thing is that appeals to brain differences feature regularly in arguments concerning self-control or moral understanding: in the case of addiction, people appeal to changes in the brain's reward system (Baler and Volkow 2006); in the case of psychopathy, they cite amygdala dysfunction as a factor that reduces responsibility (Glannon 2008). I will follow current practice and talk about such differences as dysfunctions, but one could in principle recast the argument in terms of brain differences that underlie psychological dysfunction. For the discussion about moral responsibility, what matters is the relationship between brain difference and psychological dysfunction, not whether we call this difference dysfunctional at the level of the brain. But it is important to recognize that not all brain differences or defects will ground mental dysfunction, and some mental dysfunctions will not be relevant for responsibility, because they affect areas of cognition and perception which are not relevant to moral judgment and decision-making (Jefferson and Sifferd 2018).

3.1.3 The Relation between Brain Pathology, Psychological Dysfunction and Moral Responsibility

When do brain differences excuse? It needs to be the case that a certain type of brain difference realizes or causes a psychological problem that leads to diminished or absent responsibility. The relation between psychological dysfunction, brain pathology and moral responsibility can play out in (at least) three possible ways.

First, we could have psychological dysfunction which is relevant to moral responsibility but not associated with brain pathology. Assume, for example, that we cannot find any type of brain difference associated with a certain psychological dysfunction, and that there is nothing systematic to be said about how that dysfunction is realized in the brain. This is not an unlikely scenario, given that mental states can be realized by many different sorts of physical states (Papineau 1994; Schramme 2013), though the extent to which this is the case is an empirical question. Currently, many scientists believe depression is unlikely to be associated with specific brain differences (Radden 2018). In such cases, whatever token brain processes realize the dysfunctional mental states leading to diminished responsibility would still be causally relevant to any lack of responsibility. However, because there is no specific type of brain pathology underlying the psychological problems and there might have been a different neurobiological basis of such mental dysfunctions in another person (or within the same person at a different time), these are cases of causal relevance without any explanatory power. There is then no *type* of brain pathology that is relevant for moral responsibility. All the explanatory work happens at the level of the mental and is based on an assessment of the individual's psychological dysfunctions.

Conversely, not every case of brain pathology will be associated with psychological problems relevant to moral responsibility. Some brain lesions, tumors, or cysts have no detectable effect at all. Others primarily affect motor functions or perceptual functions that are not relevant to moral judgment and action (e.g. motor neuron disease).

Finally, there will be cases where we can establish a relationship between specific brain dysfunction and psychological dysfunctions that disrupt capacities necessary for moral agency. For example, it has been suggested that functional differences in the amygdala and the orbitofrontal cortex of psychopaths show that they lack the empathy and impulse control necessary for full moral responsibility (Levy 2007b; Sifferd and Hirstein 2013). In order to establish whether brain pathology is relevant to moral responsibility, we thus need to (a) show that a specific type of brain dysfunction underlies a specific psychological dysfunction; and (b) show that the type of psychological dysfunction is relevant to moral responsibility and to the reasons that justify moral praise or blame for certain actions. As Nicole Vincent points out: "Neurological conditions do not undermine responsibility simply by virtue of being disorders, but rather they do so in virtue of the effect which they have on our mental capacities . . . which are required for moral agency" (Vincent 2008, 200). We may, for example, suffer from depression and low mood while still being able to recognize and react to moral reasons. The finding that someone suffers from a mental or a brain disorder needs to be supplemented by evidence that they exhibit psychological dysfunctions that are relevant to their moral responsibility, either across the board or in a specific situation.[3]

There may well be a *de facto* tendency to think that the mere presence of brain dysfunction in mental disorders shows that individuals are somehow ruled by brute physical mechanisms and are no longer reasons responsive or else lacking in control. But, as Nomy Arpaly (2005) has argued, this jump in reasoning is unjustified if there is no further evidence that the disorder has these effects. It follows that brain pathology should affect our moral responsibility judgments by informing our judgments of psychological dysfunction relevant to moral responsibility.

3.2 Brain Pathology as Evidence for Mental Dysfunction

When there is a link between specific brain anomalies and mental dysfunction, it is natural to think that our knowledge of brain dysfunction can inform our judgments of individuals' morally relevant mental dysfunctions, e.g. problems with impulse control or empathic deficits. For example, brain differences might be evidence for executive function problems, and this might be relevant for moral responsibility because an agent's ability to control their reactions or to attend to morally salient features of a situation are implicated (for example, in the context of autism or psychopathy). Indeed, brain data are currently used as evidence for psychological deficits in legal defenses (Catley and Claydon 2015). A recent review article states that "In 2012 alone, over 250 judicial opinions (. . .) cited defendants arguing in some form or another that their 'brains made them do it'" (Farahany 2016, 486).

Legal responsibility is of course not the same thing as moral responsibility, but there is overlap between the deficits that provide candidate excusing conditions in the moral and the legal realm, such as understanding what one is doing and that it is (legally or morally) wrong. Neuroscience and knowledge of brain pathology can inform our assessment of legal and of moral responsibility by giving us more information on a certain condition and its psychological profile.

This can be illustrated with reference to a class of non-pathological cases. In deciding that teenagers should not be given the death penalty, the U.S. Supreme court appeared to be influenced by brain data. The evidence

---

[3] For detailed defenses of the claim that we cannot simply move from a psychiatric diagnosis to a verdict on moral responsibility, see King and May (2018) and Jefferson and Sifferd (2018).

cited showed that teenagers' brain development is not yet complete (Sifferd 2013), and they therefore tend to have less impulse control and planning abilities than adults. If insufficiently developed impulse control decreases moral responsibility, then the brain differences are further evidence that teenagers are not responsible to the same extent as adults. Similarly, brain evidence is used in the context of psychopathy to bolster the claim that psychopaths have responsibility-relevant affective deficits and problems with impulse control (Glannon 2008; Levy 2007b). However, Stephen Morse has prominently argued that the usefulness of these data for establishing psychological deficits is limited. Regarding the case of adolescents, Morse insists that we already knew that adolescents have worse impulse control, and that the brain findings do not add anything new:

> What did the neuroscientific evidence about the juvenile brain add? It was consistent with the undeniable behavioral data and perhaps provided a partial causal explanation of the behavioral differences. The neuroscience data was therefore merely additive and only indirectly relevant to the behavioral criteria for responsibility.   (Morse 2011a, 853)

I concur with Morse on this point, but it is worth mentioning that neuroscientific evidence may have helped shift people's thinking, by showing just how deep the differences between adults and adolescents go, as Katrina Sifferd points out (personal communication). So, while the neuroscientific data *should* not have made much of a difference in this case, given what we already knew, it is plausible that they did in fact. One might object to this on the basis that the brain difference gives us additional information in the following sense: it shows us why adolescents are more impulsive, namely because their brains aren't fully developed. However, this fails to substantially change the evidential dynamic, for several reasons.

Evidence of brain pathology would rightly revolutionize our ascriptions of moral responsibility if knowledge of brain pathology could give us insights into mental deficits independently of what we could have gleaned through behavior or self-report. However, the way we currently establish the relationship between psychological dysfunction and brain difference makes this goal unrealistic for the time being. To find out what constitutes normal and abnormal brain function, for example via imaging techniques, neuroscientists need to correlate brain data with clearly delineated psychological phenomena. In trying to establish, for example, which brain differences (if any) correlate with deficits in executive function, we need a clear psychological test for normal executive function. One can then look at brain function during tasks that test executive function. For example, in the Wisconsin Card Sorting Test, participants sort cards according to rules that change throughout the task and the speed with which they adjust to this is measured. Such tasks, along with neuroimaging, can establish brain differences between people who exhibit problems in executive function and those who do not. The individuation of relevant brain areas and functions is done by averaging over groups of subjects. This means that, as Morse (2017) points out, we rely on clearly identifiable mental phenomena and processes to find out whether there are related brain differences. Thus, we already need to be able to identify a psychological deficit in order to find its brain correlate (cf. Levy 2007a, 149).

When establishing connections between data from neuroscience and psychological processes, neuroscientists need to navigate some well-known methodological pitfalls. Take for instance the process of reverse inference, where scientists attempt to draw conclusions regarding likely psychological processes from brain data, by appealing to known correlations between activation of a certain brain area and a certain psychological process. For example, we might infer that an individual finds a certain activity rewarding because a brain region of hers that has been correlated with reward in a previous study is active. However, this inference is only strongly supported if that area is *only* implicated in reward processing and in no other psychological functions (Poldrack 2006, 2011). Neuroscientists are well aware of this problem and take this into account when assessing the reliability of the inferences they draw from brain data. A further methodological issue is that we cannot straightforwardly draw inferences about individuals on the basis of group findings, because the effect of a certain brain finding averages over the group, which may be quite heterogenous internally. Therefore, a person may be an outlier in terms of brain function while still performing normally on a psychological measure.[4]

So at least for now, knowledge of brain dysfunctions will only yield corroborative evidence for the existence of specific mental dysfunctions. Such corroborative evidence will be especially useful in cases where behavioral evidence conflicts or we are not sure whether an individual is faking a mental illness.[5] While evidence of brain difference or pathology is useful for aiding

---

[4] For a detailed discussion of problems in using brain imaging data as evidence in the legal context, see Sinnott-Armstrong et al. (2008).

[5] I thank Katrina Sifferd for pointing this out.

our understanding of mental health conditions and the way they affect reasoning and decision-making, it is only one piece in a much larger puzzle.

3.2.1 Assessing the Evidential Role—Case Studies

Where mental dysfunction mitigates responsibility and is associated with brain dysfunction, the brain dysfunction will be *a cause* of reduced responsibility because it realizes a psychological dysfunction. But even in cases where certain psychological dysfunctions are caused by, or are expressive of, brain difference, we *assess* individuals' moral responsibility at the psychological level. So, on an evidential level, the contribution of brain pathology is to corroborate the existence of psychological deficits or indicate that a certain individual is a likely candidate for a certain psychological dysfunction. This evidential role can, however, in practice be very important, as I will show below.

In the introduction, we encountered the example of Billy, the young criminal with the slow growing brain tumor. In the thought experiment, Levy stipulated that the tumor was the cause of Billy's aggressive and immoral behavior. In light of the above discussion about the evidential role of brain dysfunction, we might speculate that when an individual has a tumor of the location and size that Billy has, we can infer that their capacity for impulse control is compromised to such an extent that we could not expect the individual to be able to control their desires and behave morally. This would be an instance where brain pathology both causes and provides evidence for psychological dysfunction.

However, in real life cases, it is not straightforward to move from the knowledge that brain pathology caused certain psychological deficits to the claim that a person was not responsible, even if we know that certain types of brain pathology are normally associated with specific psychological problems. For example, we know that frontotemporal dementia is frequently associated with disinhibition (Zamboni et al. 2008). But moving from the knowledge that an individual has a brain disorder associated with specific psychological problems to the claim that they are therefore not fully responsible requires us to know the extent of the psychological deficits a person has *at the time* for which we are considering their responsibility.

Consider two real-life cases which illustrate this problem. First, Burns and Swerdlow (2003) describe the case of a man, who, following Morse (2011b), I will refer to as Mr. Oft (presumably short for "orbito-frontal tumor"):

At age 40, Mr. Oft develops an interest in child pornography, starts secretly collecting it and making sexual advances to his pre-pubescent stepdaughter. After this is revealed to his wife, the child's mother, the man is removed from the home and takes part in an inpatient rehabilitation program as a condition for not being imprisoned for his actions. However, he does not successfully complete the rehabilitation program: "Despite his strong desire to avoid prison, he could not restrain himself from soliciting sexual favors from staff and other clients at the rehabilitation center and was expelled" (Burns and Swerdlow 2003, 437). Just before going to prison, Mr. Oft is admitted to hospital with a bad headache and undergoes neurological examination because of reported balance problems. During his examination, he solicits favors from female medical staff and is unconcerned by the fact that he urinates on himself. He also shows impairment in motor tasks, such as writing and drawing a clock face. An MRI scan reveals a large right orbitofrontal tumor.

After the tumor is removed, both Mr. Oft's motor control and his behavior return to normal, and he returns to his family. Nine months later, the tumor regrows and headaches and interest in pornography recur. The tumor is once again operated upon. Burns and Swerdlow claim that because orbitofrontal lesions led to a loss of impulse control, the patient "could not refrain from acting on his pedophilia despite the awareness that this behavior was inappropriate" (Burns and Swerdlow 2003), implying that the man was not responsible for his actions.

A further, less extreme, example which does not involve criminal behavior is that of brain scientist Barbara Lipska, who suffered from numerous brain tumors which were successfully treated. She subsequently describes the changes to her personality that these caused:

> I didn't suddenly become someone else. Rather, some of my normal traits and behaviors became exaggerated and distorted, as if I were turning into a caricature of myself. (...) I had no time for anything — not even for the things that I really enjoyed, like talking to my children and my sister on the telephone. I would cut them off midsentence, running somewhere to do something of great importance, though what exactly, I couldn't say. I became rude, and snapped at anyone who threatened to distract me. (...) Strangely, I wasn't worried. Like so many patients with mental illness, whose brains I had studied for a lifetime, I was losing my grasp on reality.
>
> (Lipska 2016)

As in the case of Mr. Oft, Lipska's behavior changed back to normal once the tumors were successfully treated. It is clear that the tumors were causally involved in the behavioral changes both individuals experienced; it is also clear that the tumors corroborate the existence of psychological dysfunction that manifests itself in problematic behavior. It is less clear that knowing about the tumors helps us to decide at what stage, if any, the impairments were so substantial that moral responsibility was lost or reduced.

The case of Mr. Oft illustrates both the limits and the importance of brain data for assessing moral responsibility. While the brain tumor affected his control, it is possible that he was still fully responsible when he molested his stepdaughter, because control had not yet deteriorated significantly. Looking at a brain scan will not enable us to tell what level of impulse control Mr. Oft had, even when it is clear that his brain abnormality was causally responsible for his behavioral changes.

However, knowing there is a brain tumor which affects areas associated with executive function is still relevant. As Sifferd (2013) points out, absent information about brain pathology, Mr. Oft's behavior is far more difficult to place as showing problems in impulse control, because behavior such as seeking out sexual favors and collecting pornography, and even molesting one's stepdaughter, may occur in the absence of impulse control problems. It is important to bear in mind that despite quite erratic behavior, Mr. Oft would have been imprisoned, had he not been admitted to hospital because of his headache and received neurological testing because of his balance problems. In other words, a tumor that did not cause the same kind of physical symptoms—headaches, balance problems—would likely have been discovered later and physicians would not have explored the possibility of psychological dysfunction in the same way. This means that, practically speaking, discovering brain tumors is sometimes of immense relevance to assessing the patient's psychological capacity correctly.[6]

To summarize, knowledge of Mr. Oft's tumor is crucial in establishing the causal origin of problems with impulse control and supports the psychological diagnosis, in some cases providing the trigger to consider the possibility of psychological dysfunction. Sometimes, brain pathology provides evidence that an individual is exempted or that their moral responsibility is reduced, because it causes relevant psychological deficits. Furthermore, knowing of this brain pathology will help us place troubling behavior in context and

---

[6] Thanks to Jan-Hendrik Heinrichs for pressing this point.

confirm the kind of dysfunction at issue. But important questions will remain unanswered. Knowing that Mr. Oft suffers from a brain tumor will not tell us at what stage he can no longer be said to be responsible, so it won't help us retrospectively answer the question whether Mr. Oft was responsible when he made advances on his step-daughter.

3.3 The Causal Path and the Relevance of Changes in Psychology

3.3.1 Lack of Control over Acquiring Immoral Dispositions and Constitutive Moral Luck

We have seen that the existence of brain pathology can provide important information relevant to responsibility in cases where there is a known link between brain pathology and psychological functioning. However, there is a common intuition that causal history also matters to responsibility, and that the causal history of people with brain pathology can (partially) excuse. In his example of Billy, the young criminal with a brain tumor, Levy argues that the reason we should not blame Billy is because the causal path of how people acquire immoral dispositions matters to responsibility: "The agent's causal history matters crucially to our assessment of his responsibility" (Levy 2007b, 134). The underlying idea seems to be that we are not responsible for moral deficits acquired through physical illness. On Levy's proposal, it is not just synchronic mental dysfunction that matters, but also the causal path that led to this dysfunction.

The idea that causal histories matter is a recurring theme in the philosophy of moral responsibility. In the context of brain pathology, the question that arises is whether the fact that an individual had no control over acquiring morally problematic characteristics excuses (partially or wholly). I will argue against the claim that causal histories mitigate responsibility when people are not in control of their personality changes. However, there is an important way in which personality changes do matter to moral responsibility, because they undermine individuals' established ecological mechanisms of gaining moral knowledge and controlling their behavior, as I show in the next section.

If causal history is to matter to moral responsibility, then we should not focus on cases where a person suffering from brain pathology clearly lacks impulse control and an understanding of their action. In those cases, agents fail to meet criteria for full responsibility at the time of action and an excuse to causal history becomes unnecessary. While the causal history might contribute to the explanation of current deficits, the existence of current deficits is sufficient to excuse or mitigate responsibility.[7]

If causal history matters, it needs to add something. So the interesting cases are those where, looking at the individual's present psychological capacities, we would not judge them incompetent, but the causal history involving a brain pathology provides reasons to excuse them. For example, Mr. Oft may or may not have had pedophilic urges from the outset, but it appears that they might have become stronger or that his impulse control got worse. It is likely that there will have been stages of the brain disease where moral behavior became harder, but not prohibitively hard for the affected person. They would at this later time, call it t_2, have the capacities we normally take to be sufficient for moral responsibility, even though they would not have the same capacities that they had before they became ill, at t_1. In other words, the individual at t_1 and t_2 would both be considered responsible if you only look at the relevant time slice. Even though the one at t_2 does not have equally well-developed moral reasons-responsiveness as the one at t_1 (they might be more compulsive, less able to concentrate or more prone to anger), they both pass a threshold of responsiveness necessary for moral responsibility. (We will elegantly pass over the question where exactly that threshold lies.) At t_2, the "lack of psychological capacities for impulse control" excuse would not have applied if we just looked at the person's capacities at t_2, and the person should be counted as responsible by synchronic criteria. However, they might count as having a (partial) excuse because of the way they came to the capacities they have at t_2.

Walter Sinnott-Armstrong explicitly endorses the claim that "sometimes, the tumor adds force to the excuse by raising the threshold of control required for responsibility" (Sinnott-Armstrong 2012, 203). So even people whose capacities would normally count as good enough for being responsible, may no longer count as responsible. He points to the fact that the idea that causation matters appears in the Model Penal Code as well, which

---

[7] There are exceptions to the rule that current incapacity excuses. These are discussed is the debate about "tracing," which concerns the question how to best to deal with cases where cause harm while temporarily lacking self-control (for example, because they are drunk or high on drugs), but are responsible for voluntarily getting themselves into the current incapacitated state in the first place (see King 2014 for discussion). These cases are the converse of the brain disorder cases I discuss below—rather than the causal path to current capacity being thought to reduce responsibility as in the brain tumor cases, in the tracing cases it increases it beyond the responsibility we would attribute on synchronous criteria alone.

specifies that mental incapacity must be caused by a mental defect or disease. On this reasoning, if the cause of the reduced capacity is somehow external to the agent and beyond their control, this changes the threshold for responsibility. John Martin Fischer and Mark Ravizza, too, make the general claim that responsibility is a historical notion and that it is not just the current properties of an agent, but the way these were acquired, that decide whether the agent is morally responsible (Fischer and Ravizza 1998, 187)— an agent may look responsible by synchronic standards, but not be responsible because of the way they acquired their capacities. So the proposal to be considered is that historical facts about the way individuals came to have their current level of reasons responsiveness also matter to whether they are fully responsible.

Should we accept the claim that an agent who is responsible by synchronic criteria at t_2 is not in fact fully responsible because they are not responsible for acquiring the psychological features that make them behave badly, because these resulted from brain pathology? Cases where agents are responsible by synchronic standards but have not acquired their current values and desires in the normal way are discussed in the context of manipulation scenarios, in which a nefarious person manipulates an individual's psychology to achieve his own goals. Some authors deny that such manipulated individuals are responsible, even if they meet synchronic or current time-slice criteria for responsibility (Fischer and Ravizza 1998). Manipulation scenarios and cases of brain tumors or neuro-degenerative diseases share the feature that the agent's psychology is changed through brute processes which are external to their agency and non-transparent to them.[8]

One way of denying history sensitivity for these cases is to point out that everybody's moral character is shaped by many factors that are outside their control. We are all subject to constitutive moral luck in that, to a large extent, we do not control the environment and the genetic material that shape our personality. Our influence on the traits which lead us to behave in certain ways is extremely limited. Some people develop a bad character and are deficient in their ability to understand and respond to moral reasons because of their horrible parents, or because they never had a chance to learn

---

[8] Manipulation cases also have the added feature that there is another agent responsible for the psychological changes in the subject, which may well contribute further to changes in our responsibility intuitions.

certain values. Brain tumors which change our personality for the worse are just a more extreme case of this problem.[9]

But maybe there is something special about brain pathology that goes beyond the way luck normally affects our character and dispositions? Maybe some histories are special. The most obvious justification for such a claim would be that in the case of brain pathology, a person who used to have an intact moral character develops a problematic one because of a physical disease, without any voluntary contribution or moral slippage through embracing bad habits.[10] However, falling into bad habits is likely caused by a combination of having certain predispositions which are determined by nature and nurture, and the circumstances one happens to be in. Just as one can be unlucky in being placed in a morally challenging environment, such as a totalitarian state, one can be unlucky in acquiring deviant impulses through a disorder. Similarly, the fact that someone is a pedophile and sexually attracted to children does not mean that they are not responsible for acting on this attraction, even though they may have no control over feeling this attraction in the first place, and they are certainly unlucky to have that disposition. If they have acquired normal self-control and reflection abilities, we can expect them to use these. The same applies to people who are naturally stingy, short-tempered, etc. So an appeal to the fact that agents had no control over acquiring certain dispositions will not help to distinguish between constitutive moral luck that we should accept and luck that undermines responsibility. In fact, the perceived impossibility of drawing a line between acceptable and unacceptable forms of moral luck is a reason why some authors argue that history sensitive compatibilism is not a tenable position (Levy 2009; Arpaly 2002).

3.3.2 The Effect of Psychological Change on Morally Responsible Agency

Lack of control over personality changes is not a good candidate mitigating condition. However, there is an important related phenomenon that does affect the psychological capacities necessary for moral responsibility and

---

[9] One might of course be tempted to draw the opposite conclusion and take this kind of case as a reason to reject responsibility altogether, as so many relevant factors are ultimately out of our control. I take this to be an unattractive option and will not pursue it further.

[10] In fact, personality change plays a major role in some arguments for the mitigated responsibility of individuals with brain tumors, cf. Reznek1997 and Sinnott Armstrong 2012.

thereby provides some excuse. I believe that Sinnott-Armstrong is right to say that we are missing something important if we only look at psychological features such as self-control or reasoning. But what supplies an excuse in these cases is something different that hasn't been appreciated in the literature: what matters is how the *change* in capacities influences synchronic capacities, some of which rely on support from the social environment. Let me explain.

Individuals suffering from brain pathology started out with normally developed moral capacities, and these change because the illness affects their interests and concerns, impulse control, and moral judgment, as well as their awareness of these changes. In her memoir on living with a brain tumor, Lipska (2018) vividly describes her lack of insight into the changes in her personality, as well as the fact that her family felt overwhelmed and did not know how to deal with the new behavior. She became irritable, impulsive, and demanding but did not notice the way these changes were distressing her family. Her family, in turn, was not able to call out the challenging behavior because they were unable to cope with the changes in someone who used to be considerate. It is plausible that over and above the impairment of impulse control, Lipska's responsible agency was affected by the fact that (a) she was unaware of the changes in her behavior and attitudes, (b) she hadn't developed mechanisms to cope with the changes, and that (c) her family was too shocked and concerned by the changes to hold her accountable. Both she and her environment were insufficiently equipped to deal with the changes in her psychology due to her brain disease, and this further affects responsibility.

These problems can be illustrated more clearly with some further examples. As adults with developed moral capacities, we know our moral weaknesses and try to counterbalance them. Let us assume that Clara knows that she has a slightly flighty nature and will be unfaithful to her partner if separated from him for extended periods of time. In order to avoid hurting him and her relationship, she therefore avoids long periods of separation. If, due to a brain tumor or a neurodegenerative condition, she all of a sudden finds her sexual urges much stronger, the measures she has put in place to remain faithful may well be insufficient. Furthermore, if she does not realize that there has been a drastic change in her psychology, she will not see the need to take steps to adjust to her stronger urges. We could speculate that something similar might have occurred in the case of Mr. Oft: he had acquired ways of dealing with unwelcome impulses of a certain strength through distraction, as part of his moral education. When the impulses became stronger due to the tumor, he was less able to control his behavior, because he hadn't learned to cope with impulses of that strength. In these ways, certain brain disorders can mitigate responsibility because such pathologies can present one with new challenges that one has not developed strategies for. Arguably, one of the reasons we hold children and teenagers to a different standard of responsibility is not just that they have less raw impulse control, they also have had less time to get to know themselves and develop ways to avoid temptation. They have not yet been taught to count to ten when they are angry, or to check their knee jerk reaction with a sympathetic interlocutor to see whether they are overreacting.

As can be seen especially vividly in the example of young people, our moral responsibility is socially supported or "scaffolded." McGeer and Pettit (2015) argue that other people's actual and imagined reactions to our moral and immoral actions play a role both in motivating us to do the right thing and in establishing, in a collaborative effort, what the right thing to do is. The extent to which our social environment contributes to our moral agency, in supporting both our sensitivity to moral reasons and our motivation to act on them, is increasingly recognized in the literature (Washington and Kelly 2016; Holroyd 2018). McGeer and Pettit make the plausible claim that part of what motivates us to behave morally is that we want to be able to justify our behavior to others. The fact that others call us out when we do wrong is one of the factors that keeps us on the straight and narrow. Our moral capacities are developed and scaffolded through the way we are embedded into a community that reacts to our moral and immoral behavior through praise, blame, punishment, etc. If an adult loses developed moral capacities, it will be harder for others to calibrate their responses to that person, which in turn makes it even more difficult for that person to respond to moral reasons. When people are baffled by our changed conduct, they sometimes do not call out inappropriate behavior, because they are at a loss how to respond to out-of-character behavior. This effect is likely to be aggravated if we *know* that the person suffers from a serious illness and the new behavior is caused by that illness.

Furthermore, just as we can make it easier to behave well by avoiding certain situations where we are tempted to behave badly in normal circumstances, we can enlist the help of others more directly in behaving well. We can get others to remind us to do things we know we should do, tell us not to send that e-mail to our obnoxious relative before having a night to cool down, etc. But, for this to work, the people who take on these roles in our social environment need to know what our weak spots are and *how* weak these are. So when our personality changes due to illness, these important relational factors supporting our responsible agency are endangered as well.

We need not adopt a history-sensitive view of moral responsibility in order to accept the claim that the psychological changes in some brain disorders provide a mitigating factor. What makes responsibility harder for those individuals who suffer from progressive brain pathology is the fact that their self-control, self-monitoring, and social support networks are not equipped to deal with the personality changes. The correct description would not be that somebody who was responsible by synchronic criteria while ill is not responsible because of the way they became ill. Rather, the problem arises from an overly narrow focus purely on intrinsic measures such as the strength of certain impulses or self-control (executive functions), even synchronically speaking. We need to also look at habits and mechanisms (both internal and external in terms of social scaffolding) that a person has developed in order to deal with familiar challenges and at the way these are supported by their social environment.

Washington and Kelly make a related point when discussing moral responsibility for implicit bias. In becoming aware of and controlling for implicit bias, we are heavily dependent on expert knowledge in our environment and on indirect measures to control it. They draw an analogy to blood pressure: "Moreover, no one can directly control her own blood pressure, or bring about an immediate and sustained change in it by direct act of will. To effectively control our blood pressure, most of us need to learn about and use the more roundabout, external methods that have been empirically verified" (Washington and Kelly 2016, 27–28). Just as we need to put external checks and indirect measures in place to monitor and control our blood pressure, we need to put indirect measures in place to assure self-control and sensitivity to moral reasons for action.

Let me give another example. Some people are prone to losing their temper easily when hungry. If they know this and their partner knows this, they will make sure that snacks are in supply at the right time. But if a newly acquired health condition means that this threshold for hunger-induced irritability has moved, your partner will not know when to bring out the cookies to keep you on an even keel and prevent intemperate bursts of rage. It is the fact that this system of support, accountability, and control extending over more than just the individual has been disrupted that can provide further excuse or mitigation in the case of progressive brain diseases.

Importantly, this kind of mitigating or excusing condition does not only apply to paradigmatic brain diseases such as tumors. Similar considerations apply in other mental disorders, for example schizophrenia or psychosis. If individuals who suffer from these conditions start having unusual experiences and they and their social environment have not yet developed ways of categorizing and coping with them, this reduces responsibility. In disorders such as schizophrenia, social support of responsible agency may be further hindered by the fact that friends and family do not understand the nature and scope of the condition, and therefore find it hard to know how to tailor their ways of holding the affected individual responsible. While brain tumors provide a particularly vivid example and the progression is different, similar considerations can apply when individuals develop mental disorders like schizophrenia or bipolar disorder.

3.4 Conclusion

I have argued that the primary role brain dysfunction or pathology should play in our practices of blame is to provide further evidence and explanation for responsibility-relevant psychological dysfunction. Translating brain data into relevant psychological dysfunction is not always straightforward. It is particularly difficult when a patient has a progressive brain disease and we are trying to establish the point where psychological functioning is sufficiently impaired for the patient's responsibility to be diminished or lost. I have considered whether the causal path by which people come to behave badly when they suffer from brain disorders matters to their responsibility. Specifically, I have considered whether the bar for excuse is lowered by the fact that the moral deficits are a result of brain pathology. I conclude that it is not the causal path that diminishes responsibility, but the way in which psychological changes undermine the normal mechanisms of control and moral feedback.[11]

References

Arpaly, Nomy. 2002. *Unprincipled Virtue: An Inquiry into Moral Agency.* Oxford: Oxford University Press.

[11] I would like to acknowledge support by the Leverhulme Trust through an Early Career Fellowship entitled "Mental Disorders, Brain Disorders and Moral Responsibility." This chapter was presented at the work in progress seminar at the University of Birmingham (UK) and at the workshop "Responsibility and Mental Disorder" at the University of Alabama at Birmingham; and I received insightful comments from both audiences. I thank Joshua May, Matt King, Katrina Sifferd, and Jan-Hendrik Heinrichs for helpful feedback on earlier drafts of this chapter.

Arpaly, Nomy. 2005. "How It Is Not 'Just Like Diabetes'. Mental Disorders and the Moral Psychologist." *Philosophical Issues* 15: 282–298.

Baler, R. D., and N. D. Volkow. 2006. "Drug Addiction: The Neurobiology of Disrupted Self-Control." *Trends Mol Med* 12 (12): 559–566. doi: 10.1016/j.molmed.2006.10.005.

Brink, David O., and Dana K. Nelkin. 2013. "Fairness and the Architecture of Responsibility." *Oxford Studies in Agency and Responsibility* 1: 284–313.

Broome, Matthew R., Lisa Bortolotti, and Matteo Mameli. 2010. "Moral Responsibility and Mental Illness: A Case Study." *Cambridge Quarterly of Healthcare Ethics* 19 (2): 179–187. doi: 10.1017/S0963180109990442.

Burns, Jeffrey M., and Russell H. Swerdlow. 2003. "Right Orbitofrontal Tumor with Pedophilia Symptom and Constructional Apraxia Sign." *Archives of Neurology* 60 (3): 437–440. doi: 10.1001/archneur.60.3.437.

Catley, Paul, and Lisa Claydon. 2015. "The Use of Neuroscientific Evidence in the Courtroom by Those Accused of Criminal Offenses in England and Wales." *Journal of Law and the Biosciences* 2 (3): 510–549. doi: 10.1093/jlb/lsv025.

Cuthbert, Bruce N. 2014. "The RDoC Framework: Facilitating Transition from ICD/DSM to Dimensional Approaches that Integrate Neuroscience and Psychopathology." *World Psychiatry* 13 (1): 28–35. doi: 10.1002/wps.20087.

Farahany, Nita A. 2016. "Neuroscience and Behavioral Genetics in US Criminal Law: An Empirical Analysis." *Journal of Law and the Biosciences* 2 (3): 485–509. doi: 10.1093/jlb/lsv059.

Fine, Cordelia, and Jeanette Kennett. 2004. "Mental Impairment, Moral Understanding and Criminal Responsibility: Psychopathy and the Purposes of Punishment." *International Journal of Law and Psychiatry* 27: 425–443.

Fischer, John Martin, and Mark Ravizza. 1998. *Responsibility and Control: A Theory of Moral Responsibility.* Cambridge: Cambridge University Press.

Frankfurt, Harry G. 1971. "Freedom of the Will and the Concept of a Person." *The Journal of Philosophy* 68 (1): 5–20. doi: 10.2307/2024717.

Glannon, Walter. 2008. "Moral Responsibility and the Psychopath." *Neuroethics* 1 (3).

Graham, George. 2013. *The Disordered Mind: An Introduction to Philosophy of Mind and Mental Illness.* New York: Routledge.

Holroyd, Jules. 2018. "Two Ways of Socializing Moral Responsibility: Oppression, Politics and Moral Ecology." In *Social Dimensions of Moral Responsibility*, edited by Katrina Hutchison, Catriona Mackenzie, and Marina Oshana, 137–162. Oxford: Oxford University Press.

Insel, Thomas. 2013. "Transforming Diagnosis." NIMH, accessed 07.06.2017. https://www.nimh.nih.gov/about/directors/thomas-insel/blog/2013/trans forming-diagnosis.shtml.

Jefferson, Anneli. 2020. "What Does It Take to Be a Brain Disorder?" *Synthese* 197 (1): 249–262. doi: 10.1007/s11229-018-1784-x.

Jefferson, Anneli, and Katrina Sifferd. 2018. "Are Psychopaths Legally Insane?" *European Journal of Analytic Philosophy*14 (1): 79–96.

King, Matt. 2014. "Traction without Tracing: A (Partial) Solution for Control-Based Accounts of Moral Responsibility." *European Journal of Philosophy* 22 (3): 463–482. doi: 10.1111/j.1468–0378.2011.00502.x.

King, Matt, and Joshua May. 2018. "Moral Responsibility and Mental Illness: A Call for Nuance." *Neuroethics* 11 (1): 11–22. doi: 10.1007/s12152-017-9345-4.

Levy, Neil. 2007a. *Neuroethics, Challenges for the 21st Century.* Cambridge: Cambridge University Press.

Levy, Neil. 2007b. "The Responsibility of the Psychopath Revisited." *Philosophy, Psychiatry, and Psychology* 14 (2): pp. 129–138.

Levy, Neil. 2009. "Luck and History-Sensitive Compatibilism." *The Philosophical Quarterly* 59 (235): 237–251. doi: 10.1111/j.1467–9213.2008.568.x.

Lipska, Barbara. 2016. "The Neuroscientist Who Lost Her Mind." *The New York Times.* Accessed 14.2.2019. https://www.nytimes.com/2016/03/13/opinion/sunday/the-neuroscientist-who-lost-her-mind.html.

Lipska, Barbara. 2018. *The Neuroscientist Who Lost Her Mind: A Memoir of Madness and Recovery.* New York: Bantam Press.

McGeer, Victoria, and Philip Pettit. 2015. "The Hard Problem of Responsibility." In *Oxford Studies in Agency and Responsibility.* Oxford: Oxford University Press.

Morse, Stephen. 2011a. "Avoiding Irrational Neurolaw Exuberance: A Plea for Neuromodesty." *Mercer Law Review* 62: 837–859.

Morse, Stephen. 2011b. "Lost in Translation? An Essay on Law and Neuroscience." In *Law and Neuroscience: Current Legal Issues Volume 13,* edited by Michael Freeman, 529–562. Oxford: Oxford University Press.

Morse, Stephen J. 2017. "Neuroethics: Neurolaw." In *Oxford Handbooks Online.* doi: 10.1093/oxfordhb/9780199935314.013.45.

Papineau, David. 1994. "Mental Disorder, Illness and Biological Disfunction." *Royal Institute of Philosophy Supplements* 37: 73–82. doi: 10.1017/S135824610000998X.

Pickard, Hanna. 2015. "Psychopathology and the Ability to Do Otherwise." *Philosophy and Phenomenological Research* 90 (1): 135–163. doi: 10.1111/phpr.12025.

Poldrack, Russell A. 2006. "Can Cognitive Processes Be Inferred from Neuroimaging Data?" *Trends in Cognitive Sciences* 10 (2): 59–63. doi: https://doi.org/10.1016/j.tics.2005.12.004.

Poldrack, Russell A. 2011. "Inferring Mental States from Neuroimaging Data: From Reverse Inference to Large-Scale Decoding." *Neuron* 72 (5): 692–697. doi: 10.1016/j.neuron.2011.11.001.

Radden, Jennifer. 2018. "Rethinking Disease in Psychiatry: Disease Models and the Medical Imaginary." *Journal of Evaluation in Clinical Practice* 24 (5): 1087–1092. doi: 10.1111/jep.12982.

Reznek, Lawrie. 1997. *Evil or Ill? Justifying the Insanity Defence.* London: Routledge.

Schramme, Thomas. 2013. "On the Autonomy of the Concept of Disease in Psychiatry." *Frontiers in Psychology* 4 (457). doi: 10.3389/fpsyg.2013.00457.

Shoemaker, David W. 2011. "Psychopathy, Responsibility, and the Moral/Conventional Distinction." *The Southern Journal of Philosophy* 49: 99–124. doi: 10.1111/j.2041-6962.2011.00060.x.

Sifferd, Katrina. 2013. "Translating Scientific Evidence into the Language of the 'Folk': Executive Function as Capacity-Responsibility." In *Legal Responsibility and Neuroscience*, edited by Nicole A. Vincent. Oxford: Oxford University Press.

Sifferd, Katrina, and William Hirstein. 2013. "On the Criminal Culpability of Successful and Unsuccessful Psychopaths." *Neuroethics* 6 (1):129–140.

Sinnott-Armstrong, Walter. 2012. "A Case Study in Neuroscience and Responsibility." *Nomos* 52: 194–211.

Sinnott-Armstrong, Walter, Adina Roskies, Teneille Brown, and Emily Murphy. 2008. "Brain Images as Legal Evidence." *Episteme* 5 (3): 359–373. doi: 10.3366/E1742360008000452.

Vincent, Nicole A. 2008. "Responsibility, Dysfunction and Capacity." *Neuroethics* 1 (3): 199–204. doi: 10.1007/s12152-008-9022-8.

Washington, Natalia, and Daniel Kelly. 2016. "Who's Responsible for This?: Moral Responsibility, Externalism, and Knowledge about Implicit Bias." In *Implicit Bias and Philosophy, Volume 2.* Oxford: Oxford University Press.

Zamboni, G., E. D. Huey, F. Krueger, P. F. Nichelli, and J. Grafman. 2008. "Apathy and Disinhibition in Frontotemporal Dementia: Insights into Their Neural Correlates." *Neurology* 71 (10): 736–742. doi: 10.1212/01.wnl.0000324920.96835.95.

4

Taking Control with Mechanisms
of Psychotherapy

Robyn Repko Waller

Are psychiatric patients responsible agents? Recent accounts of moral responsibility have aimed to capture in what sense agents with particular disorders have, or lack, agential control and moral standing. In this chapter I address this philosophical question through the lens of effective psycho-therapy, specifically cognitive-behavioral approaches. I argue that a course of successful talk therapy brings individuals with disorders of agency from initial marginal agency to closer to full-blown agency. I further argue that the therapeutic mechanisms, or techniques, that bring about this change in moral agency do so in part via enhancing patients' agential control capa-cities, in addition to effects on patients' affect and values. The primacy of control in explanations of psychopathology, as well as in underscoring excuses and exemptions of such agents from select moral practices, is best illustrated via how patients' responsiveness to reasons changes over the course of therapy.

In Section 4.1, I introduce what kinds of agential capacities underscore responsible agency and how agents with psychological disorders are included in, or exempted from, our moral practices. In Section 4.2, I canvass a proposed psychiatric explanatory construct of psychopathology, emotion regulation, and related techniques of agential change in talk ther-apy. In Sections 4.3–4.6, I explore to what extent changes in agential capacities over a course of successful talk therapy are control-related. Specifically, I consider the control capacities of such patients pre-therapy in light of behavior and then in light of therapeutic goals. I make the case for this account of the agency of individuals with disorders of agency by applying it, first, to agoraphobia and Exposure Therapy and, second, to borderline personality disorder and Dialectical Behavior Therapy. The approach bridges philosophical concepts of reasons-receptivity and

Robyn Repko Waller, *Taking Control with Mechanisms of Psychotherapy* In: *Agency in Mental Disorder: Philosophical Dimensions.* Edited by: Matt King & Joshua May, Oxford University Press. © Robyn Repko Waller 2022.
DOI: 10.1093/oso/9780198868811.003.0005

reasons-reactivity with key psychiatric constructs, such as distress tolerance and experiential avoidance.

4.1 Moral Responsibility and Disorders of Agency

We hold people responsible for their actions in the absence of exculpating factors. Imagine that, driving down a bustling Main Street on a sunny Saturday, Charles' car drives over the sidewalk and careens towards the crowd of pedestrians, injuring several people and damaging the store-front of a small business. That such events are unfortunate and devastating for those people involved and those indirectly impacted is not in question. Harm, physical and psychological, was caused to many individuals. Affected individuals would likely feel a mix of sadness, grief, fear, and anxiety. In contrast, how we react, morally, to the driver of the vehicle is dependent on further facts: First, consider the case in which Charles knowingly drove his vehicle onto the sidewalk, with the intention to maim pedestrians and to destroy property. Those individuals impacted personally at the scene should be permitted to express, and likely would express, blame in the form of moral anger and resentment for the injury and damage done. Observers and those who later hear about the incident may express blame in the form of indignation and moral outrage.[1]

Now, consider the case in which we come to find out that Charles has previously undiagnosed severe dementia and that this has contributed to his unsafe driving. Most people would soften their moral stance towards Charles here. Charles as an individual suffering from severe dementia shouldn't be to blame for these outcomes, regardless of how morally bad and unfortunate the outcome is.

The practice of blaming an agent for her morally bad actions, manifest in our reactions of moral anger, resentment, and indignation, is part of the familiar moral practices we as interconnected individuals engage in, sometimes unreflectively. We often, though without the same verve, participate in the practice of praising or crediting agents for their morally good actions. We express our gratitude and admiration to agents who act in morally good

---

[1] Perhaps an example of attempted vehicular homicide is so extreme that one might suspect there must be an exculpating factor. Charles must suffer from some condition or deficits that contributed to his morally bad decision and action. If the reader is of this mindset, then substitute a slight or injury that agents more frequently commit—an instance of verbal abuse or a physical shove.

ways (Strawson 1974). For instance, we admire individuals who stand up and speak against social injustice. Moreover, excusing agents for some action due to factors beyond their control or what they could not foresee is part and parcel of that system of moral practices. So, too, is our judgment that this agent just isn't the kind of agent we ought to blame for a whole host of their actions—Charles as an individual with severe dementia, young children, etc. Mental disorder that impacts agency plausibly falls under the category of a condition that either excuses or exempts the agent from certain kinds of moral judgment.

On such a view of moral responsibility, agents whose actions are driven by mental disorder are not held responsible because they lack, at a time or unconditionally, the agential capacities for responsible agency (e.g., Wallace 1994; Fischer and Ravizza 1998; McKenna 2012). What are the agential capacities constitutive of being a responsible agent? Here accounts of moral agency have varied, encompassing states and capacities such as the structure and proper ordering of one's motivational states (Frankfurt 1971), one's psychology being grounded in reality (Wolf 1987; Fischer and Ravizza 1998), ownership of one's motivational states and agency (Fischer and Ravizza 1998), a moral evaluational stance (Watson 1975), and responsiveness or sensitivity to reasons, moral or prudential (Wolf 1990; Wallace 1994; Haji 1998; Fischer and Ravizza 1998; Fischer 2006; Nelkin 2011; Sartorio 2016). We exempt from our moral practices, or at least mitigate responsibility for, those agents who lack, to a significant degree, one or more of these capacities or attributes. As such, it is sometimes argued, we ought to take the objective stance toward such persons—they are not candidates for moral assessment for the disorder-driven actions in question, and so it would be inappropriate to, say, blame them for their morally bad actions (Strawson 1974). Recent work in philosophy of agency has dissected the differences in agential capacities and so moral standing for agents with particular disorders, including addiction, psychopathy, depression, autism, and dementia.[2]

Let's take seriously that a proper account of holding morally responsible ought to have the sensitivity to respond to the distinctive agential profiles of individuals with distinct disorders of agency. Moral responsibility has a strong conceptual tie to free will. Possession of free will is often held to be

[2] See, e.g., Levy (2013); Greenspan (2003); Charland (2004); Watson (2004); Pearce and Pickard 2009; Nadelhoffer and Sinnott-Armstrong (2013); Kiehl and Sinnott-Armstrong (2013); Watson (2013); Levy (2014); Glannon (2017); Nelkin (2017); Godman and Jefferson (2017); Jefferson and Sifferd (2018); Shoemaker (2015); King and May (2018); Summers and Sinnott-Armstrong (2019); Gorman (2020); Wonderly (forthcoming), among others.

the control requirement for moral responsibility; agents are morally responsible for their actions insofar as they exercise appropriate control over their actions and their consequences. The agential capacities discussed as filling in accounts of being responsible, such as responsiveness to reasons, are typically taken to specify that freedom-relevant agential control (see, e.g., Fischer and Ravizza 1998). Given this nuanced difference in candidacy for participation in moral practices among clinical populations, we might press, then, the following: How do agents with distinct disorders differ in *control* capacities from nonclinical populations? Are there commonalities in control deficits across diagnoses? (That is, do these control capacities underlie psychopathology transdiagnostically?) Do distinct control profiles emerge for agents with differential diagnoses (e.g., major depression, agoraphobia, borderline personality disorder, etc.)? Further, talk therapy on a plausible construal aims to take individuals who, due to their disorder-driven actions, are at the borders of functional and so of responsible agency, and bring these individuals closer to full-blown agency and, as a corollary, fuller participation in our daily moral practices. If so, does talk therapy operate over control capacities in bringing about therapeutic change in psychological functioning?

To make inroads on these questions, in the next section I take a closer look at a proposed explanatory psychiatric construct for the nature of psychopathology and related agents of change, or techniques, of effective talk therapy. I then discuss how these explanatory psychiatric constructs are suggestive of the nature of multifaceted control deficits underpinning marginal moral agency.

4.2 Therapeutic Mechanisms of Change

What accounts for the development and maintenance of complex patterns of clinical symptoms characteristic of psychological disorders? That is, what is the nature of psychopathology? One prominent unifying construct regarding psychopathology that has received empirical support is emotion regulation (Sloan et al. 2017). Emotion regulation has been proposed as underscoring psychopathology transdiagnostically—that is, as indicating dimensions that are present across a wide range of psychological disorders that, when treated, lead to measurable clinical change. The clinical psychology literature lacks an univocal understanding of what processes and states constitute emotion regulation, but one core feature of emotion regulation frameworks is that they identify "a heterogeneous set of processes involved in modifying emotional experiences" (Sloan et al. 2017, 142). Here, by emotional experiences, clinicians refer to the web of feelings, behaviors, and constitutive physiological responses that make up or are tied to emotion. Emotion regulation is understood as a complex of strategies and skills that individuals use to process and react to their emotional experiences. Insofar as patients experience *emotion dysregulation*, their clinical symptoms persist. For instance, maladaptive strategies like rumination, suppression, and avoidance are strongly associated with symptoms of depression, anxiety, eating disorders, and substance use disorders (Aldao et al. 2010; Aldao and Nolen-Hoeksema 2012).

Moreover, insofar as emotion dysregulation is at the heart of psychopathology, effective methods of psychotherapy should affect a patient's emotion regulation profile. Numerous Behavior Therapies or Cognitive-Behavioral therapies, such as Behavioral Activation Treatment for Depression, Mindfulness-based Cognitive Therapy, Acceptance and Commitment Therapy, and Dialectical Behavior Therapy, are inclusive of the strategies or skills for emotion regulation. Here the goal of the therapeutic process is to inculcate adaptive responses to emotional experiences through the promotion of acceptance, problem-solving, and reappraisal (Hayes et al. 2011; Hofmann and Asmunden 2008; Gross and John 2003 respectively) and to break reliance on maladaptive ones. For instance, the therapeutic goal of acceptance without endorsement of certain problematic thoughts or the goal of increased distress tolerance have been classified as a therapeutic change operating over the patient's emotion-regulation capacities. Empirical research supports emotion regulation as a transdiagnostic explanatory construct. In a systematic review of the clinical literature, Sloan and colleagues (2017) found that, across diagnoses, patients evince a decrease in clinical symptoms with a decrease in emotion dysregulation over the course of successful psychotherapy. Reviewing treatment outcomes for patients with a wide variety of diagnoses, such as depression, anxiety, phobias, and borderline personality disorder, they found that treatment with cognitive-behavioral therapies of diverse orientations is associated with not only a decrease in disorder-specific symptoms but also a decrease in emotion dysregulation, as understood here. These findings suggest that a successful course of talk therapy works over emotion-regulation mechanisms and that those changes in emotion regulation are associated with meaningful changes in patients' daily functioning.

Our present question, however, is to what extent *control deficits* underscore the agential deficits of individuals with disorders of agency. Here, in the clinical literature, operative mechanisms are taken to work by altering maladaptive affective strategies and skills. To what extent, then, can we conceive of these therapy-induced changes in psychological functioning as control-related?

The "regulation" aspect of emotion regulation suggests that these therapists treat patients as *agents*—active participants in their psychological lives and in their social interactions—and train them to *intentionally* bring about change in their emotion experiences both intrapersonally and interpersonally. Indeed, patients' perceived control has also been evidenced transdiagnostically to predict outcomes of treatment with cognitive behavioral therapy (Gallagher et al. 2014). Hence, I now explore in what sense individuals with disorders of agency are agents, exercising control over their actions and, correspondingly, in what sense they can be held responsible for their actions. To do so, I start with the assumption that patients are intentional actors who bear moral responsibility, in some sense, for their actions in the clinical context. Then I discuss how many of the clinical treatment constructs included under the heading of emotion regulation, such as acceptance, avoidance, and distress tolerance, can be seen as operating over the agent's control capacities. The therapeutic techniques that fall under these headings, I argue, are the effective mechanisms that in part explain why pre-therapy agents are on the margins of agency and why post-therapy agents can move closer to full-blown agency. Further, I argue that these agential control capacities are more robust that those of basic intentional action. Using the framework of agential control as responsiveness to reasons, I argue that, indeed, such effective cognitive behavioral therapies are inducing substantive changes in patients' control capacities, and, furthermore, that these changes in agential control ground, both transdiagnostically and in disorder-specific ways, the transformation from borderline moral agent status to fuller inclusion in our responsibility practices for patient populations.

4.3 Deficits of Control Underpinning Agential Deficits of Psychopathology

The relevance of agential control for effective talk therapy and responsible agency in patient populations has been addressed previously. One prominent proposal addressing this topic comes from Hanna Pickard (2011, 2013, 2015, 2017), which is both theoretical and practical in nature, as she theorizes how clinicians ought to understand and react to the actions—especially challenging ones—of their patients in the therapeutic context. Pickard focuses on patients with personality disorder and addiction. She outlines the interpersonal difficulties of treating someone whose disorder is constituted by a set of personality traits that make them prone to life-threatening and relationship-endangering behaviors. These circumstances are an impediment to efficacious treatment insofar as the clinician must consider the patient an agent in order to encourage active change on the patient's end but may also, problematically, blame or resent the patient's actions in the context of the patient–clinician relationship. To simply take the objective stance, as Strawson (1974) understood it, toward patients would thwart the clinician's aim of effective therapy.[3] Pickard argues that clinicians who treat patients with personality disorder should judge that such patients are responsible (and perhaps even blameworthy) insofar as they act intentionally—a form of detached blame—without applying blame—a form of affective blame. Detached blame is, in essence, a form of cognitive blame without the "sting" of resentment. Clinicians can take this attitude practically if they focus on the psychosocial underpinnings of the disorder. Application of blame from the therapist acknowledges the patient as an agent and so encourages the patient to see themselves as an agent. Patients are encouraged to think "what I have done was up to me," where this ownership of action extends to neutral actions as well as challenging moral ones. Here, seeing patients with personality disorder and addiction as responsible is to see them, in some significant but limited capacity, as in control of their actions.

Are patients with personality disorder in control of their actions to the extent that they are fitting subjects of blame, even if just stripped-down cognitive blame? This is an important question to address if we are to advocate for blame as a fitting response to patients with personality disorder. Perhaps one might suggest that Pickard's stance is an instrumental

---

[3] "What I want to contrast is the attitude (or range of attitudes) of involvement or partic-ipation in a human relationship, on the one hand, and what might be called the objective attitude (or range of attitudes) to another human being, on the other...To adopt the objective attitude to another human being is to see him, perhaps, as an object of social policy; as a subject for what, in a wide range of sense, might be called treatment; as something certainly to be taken account, perhaps precautionary account, of; to be managed or handled or cured or trained..." (Strawson 1974, 8–9).

intentional stance and forward-looking approach to moral responsibility. What we really want to know, we might press, is if the patient with personality disorder is capable of controlling her behavior such that she is a *deserving* candidate of blame. Plain intentional action is, seemingly, too thin a notion of agency and control to ground this kind of responsibility. However, it is not necessarily the case that all impulsive actions of patients with personality disorder fall short of a substantive kind of voluntariness, a kind more closely tied to desert-based responsibility.[4] To see why, let's now consider how reasons-responsiveness, a capacity for stronger voluntary agency, captures the control required for being responsible.

Elsewhere, I have outlined a significant connection between reasons-responsiveness theories of control (required for moral responsibility) and therapeutic techniques for treating disorders of agency (Waller 2014). The aim in that work was to capture the boundary conditions for morally responsible agency. Specifically, I argue there that reasons-responsive accounts of moral responsibility could appropriately exempt patients with severe disorders of agency for their disorder-driven behavior and that this exemption would rely, crucially, on an understanding of the patients' control capacities pre-therapy.[5] In that context, I developed a framework for assessing, pre- and post-therapy, the reasons-responsiveness profile of such agents as an explanation of their lack of responsibility and how we might fine-tune reasons-responsive accounts to exempt such populations pre-therapy. I will use a similar approach in the present proposal.

Reasons-responsiveness accounts of moral responsibility[6] (Fischer and Ravizza 1998) specify the control capacities required for agents to be the apt targets of the moral responsibility practices. Agents must both be able to recognize a pattern of reasons for action (*reasons-receptivity*) and be able to align their conduct with these reasons because of their justificatory force (*reasons-reactivity*). Via practical reasoning mechanisms agents can exercise these capacities to various degrees, but the critical issue for Fischer and

---

[4] Ayob (2016) argues that "the mere fact that PD-impulsive behaviors are not brute bodily movements does not, by itself, provide conclusive grounds for holding that they amount to actions done voluntarily", but still "PD-impulsive behaviors are not voluntary action, at least not on a substantial reading of this term" (pp. 70–1). While I accept the first strand of Ayob's argument, I disagree with the second claim.

[5] This argument was proposed in the service of addressing an objection to Fischer and Ravizza's reasons-responsive account of moral responsibility: that some kinds of agents with severe disorders of agency (like agoraphobia) would, problematically, exhibit sufficient control to be morally responsible for disordered conduct on Fischer and Ravizza's account (Mele 2000).

[6] Here I work with the view as given in Fischer and Ravizza (1998), but see Nelkin (2011) and Sartorio (2016) for related recent reasons-responsive accounts, broadly construed.

Ravizza is whether the agent meets the standard (or threshold) of reasons-receptivity and reasons-reactivity to qualify as a *responsible* agent.

Here I follow Fischer and Ravizza (1998, 41–42) in understanding reasons as justifying reasons. A justifying reason—what Fischer and Ravizza call a "justificatorily sufficient reason"—is one that is sufficient to justify a particular action in the present circumstances, even if that reason does not move the agent to act in that way (lacks motivational sufficiency). Agents may or may not consciously entertain those reasons prior to decision and action (e.g., during deliberation). The critical point is that when there is a justifying reason R to act in a certain way given the circumstances, we can evaluate an agent's responsiveness to reasons via an evaluation of her psychology: If the agent is *receptive* to a reason R, she would recognize that reason for action—take that reason—as sufficient to justify action in those circumstances. If the agent is *reactive* to a reason R, she would act in the way indicated by that reason in those circumstances because she recognizes that reason for action. Agents can evince, then, several deficits of responsiveness to reasons: they can fail to recognize a justifying reason—a failure of reasons-receptivity with respect to that reason—or can fail to act in line with a justifying reason—a failure of reasons-reactivity with respect to that reason.

Below I demonstrate how this reasons-responsive framework for understanding agential status of individuals with disorders of agency aptly characterizes the therapeutic change, transdiagnostically, using Acceptance and Commitment Therapy as an example.[7] Acceptance and Commitment Therapy involves the central strategy of acceptance of one's thoughts, discussed in relation to emotion regulation in Section 4.2.

4.4 Responding to Reasons with Acceptance and Commitment Therapy

Acceptance and Commitment Therapy is a "third-wave" cognitive behavioral therapy. Cognitive behavioral therapy is, historically, a therapeutic

---

[7] Empirical support for control-centric reasons-responsiveness as an explanatory construct of therapeutic change would require the development of a psychometric measure of reasons-responsiveness to be used as a pre-therapy and post-therapy assessment. If talk therapy does produce psychological and behavioral changes in patients via these control capacities, then patients would evidence enhanced reasons-receptivity and/or enhanced reasons-reactivity when a course of therapy yields reduced clinical symptoms. The present project does not tackle these empirical issues, but lays the groundwork for a control-based understanding.

approach driven in part by behaviorism (sometimes termed "first wave") that was developed by psychologists such as Ellis (1969) and Beck (1976). What sets third wave cognitive behavioral therapy apart from second-wave cognitive behavioral therapies, such as Exposure Therapy, is a shift in focus from the content of one's thoughts to the relation of the agent to those thoughts (Hayes 2004). For instance, many third-wave cognitive behavioral therapies, such as Mindfulness-Based Cognitive Therapy (MBCT) (Segal, Williams, and Teasdale 2002), build in a mindfulness component to the skills taught to clients so that clients can learn to observe their emotions and mental states without working to control those emotions and states.

Acceptance and Commitment Therapy (Hayes, Strosahl, and Wilson 2011) has been demonstrated as efficacious in treating personality disorder, chronic health conditions (e.g., pain management), eating disorders, depression, anxiety, and post-traumatic stress disorder. Hence, Acceptance and Commitment Therapy seems to operate on emotion regulation capacities, transdiagnostically. Strikingly, the goal of Acceptance and Commitment Therapy is to "afford greater choice of behavior" (Gillanders et al. 2014, 83) on the part of the patient, in particular behavior that is consistent with the patient's endorsed values. Like other third-wave cognitive behavioral therapies, Acceptance and Commitment Therapy does not necessarily encourage patients to modify their emotions or cognitions but rather to change how they relate to those mental states. What marks out Acceptance and Commitment Therapy as distinct from other third-wave cognitive behavioral therapies is the disentangled goals of (a) distancing oneself from one's thoughts and *then* (b) accepting those thoughts (without necessarily endorsing or liking them). Here the core explanatory therapeutic construct that characterizes patients, such as those with personality disorder, at the initial stage of therapy is cognitive fusion. Cognitive fusion is described as "the tendency for behavior to be overly regulated and influenced by cognition" (p. 84). For instance, the Cognitive Fusion Questionnaire—a psychometric measure of cognitive fusion—contains items such as "I get so caught up in my thoughts that I am unable to do things that I most want to do" or "I tend to get very entangled in my thoughts" (Gillanders et al. 2014, 101).

The fact that patients are cognitively fused explains, according to Acceptance and Commitment Therapy proponents, the relationship between their cognition and dysfunctional action. The goal, then, of the therapeutic exercises and transformative processes of the course of therapy is to distance the client from her problematic mental states—cognitive defusion—such that she sees such states as ones that she need not act on, at least in contexts in which it is dysfunctional to do so. So the patient doesn't change the content of the thoughts in question, just their dominance in driving action. After a successful course of Acceptance and Commitment Therapy, patients are better able to recognize their thoughts as just that— "mental states and not necessarily needing to be acted upon" (Gillanders et al. 2014, 83). When thoughts are no longer taken as unconditional commands and are less dominant, the agent can exercise more flexible behaviors and so pursue a wider variety of activities consistent with her values.[8]

It is this construct of cognitive fusion that makes Acceptance and Commitment Therapy ripe for integration with a reasons-responsiveness framework. Acceptance and Commitment Therapy plausibly works on the reasons-responsive control capacities of patients to achieve therapeutic change. Let's focus, first, on the patient's reasons-receptivity. That cognitively fused patients take the content of cognitions as directing their thoughts is highly suggestive of taking those cognitions as justifying reasons for action on which patients then act. Moreover, when patients distance themselves from those cognitions through therapy, the now anemic dominance and literalness of those thoughts is indicative of a change in reasons-receptivity—that those thoughts no longer represent reasons for action. Or, perhaps, those thoughts are still taken to be reasons for action but are now weaker justifying reasons—they only justify a behavior or action in a more restricted set of circumstances. In contrast, the patient's own values, preferences, and beliefs strengthen in their role as the patient's recognized justifying reasons for action. That is, the patient retains or expands her receptivity to reasons informed by her value system.

Further, over a course of Acceptance and Commitment Therapy patients evidence a change in reasons-reactivity. The patient can now act on an altered set of recognized justifying reasons for action, informed by her values, preferences, and beliefs but not inclusive of those dysfunctional cognitions. This change is due, in part, to the lack of dominance of those previously dysfunctional cognitions. She does not just recognize her endorsed values as reasons for action but also can act more flexibly now in service of those reasons. That the patient's reasons-reactivity is enhanced is

---

[8] "...[C]entral to the construct is that the purpose of stepping back from cognitive events is to facilitate taking action that is consistent with one's values, rather than to disrupt negative thinking styles, change metacognitive beliefs or reduce stress" (Gillanders et al. 2014, 85).

what allows for her to pursue her goals, whether those are intrapersonal or interpersonal.

In this way, then, a closer analysis of Acceptance and Commitment Therapy and cognitive fusion in reasons-responsive terms promises to illuminate why certain therapeutic exercises are efficacious in alleviating dysfunctional agency and promoting behavioral flexibility. It is this freedom from dominating urges and this behavioral flexibility, the ability to act from one's own evaluational system, that is central to the control underlying moral agency and participation in the moral responsibility practices.[9] I have only offered a sketch of that analysis here. The sketch, I hope, makes a strong preliminary case for the potential marriage of reasons-responsiveness accounts of control and transdiagnostic therapeutic constructs, such as cognitive fusion, in providing a richer understanding of agency, both for clinical and nonclinical agents.

The importance of this reasons-responsive framework can be further illustrated and strengthened by applying it *to specific disorders*. Below I show how the framework can be applied to two such pairs: Exposure Therapy for agoraphobia and Dialectical Behavior Therapy for personality disorder. These two pairs provide an important contrast. Although diminished reasons-reactivity is central to both, agoraphobia involves an excessively high threshold while personality disorder involves a threshold that is abnormally low.

4.5 Agoraphobia and Exposure Therapy

Let's start with a sketch of this approach applied to agoraphobia and an efficacious course of treatment, Exposure Therapy, a "second wave" cognitive behavioral therapy.[10] The core feature of agoraphobia is an intense anxiety about being in a place or situation in which escape from or getting help in the event of panic-like symptoms[11] is difficult—e.g., being outside of one's home, being in a crowd; being in a confined place such as a bus or a

---

[9] In this way, Watson's (2004) platonic model, in which responsible agency involves acting from one's own valuational system, is also a promising candidate for integration with change in agency over therapy.

[10] This section is based on a discussion of the reasons-responsive capacities of patients with agoraphobia in Exposure Therapy in Waller (2014).

[11] As of the DSM-5, experiencing panic attacks is not a necessary condition for a patient to be diagnosed with agoraphobia.

bridge (Diagnostic and Statistical Manual of Mental Disorders 5 [DSM-5]). Panic attacks involve, typically, heart palpitations, sweating, chest pain, numbness, and a fear of death. This anxiety leads over time to avoidance of the relevant situation to avoid distress. This learned avoidance response is then reinforced whenever the patient engages in the response under threat of feared stimuli. For instance, an agoraphobic patient may seldom leave the confines of his or her home, despite realizing the missed experiences and impoverished relationships (e.g. damage to relationships and career) resulting from this behavior, due to a fear of having a panic attack.

To illustrate, consider the hypothetical example of Fred whose agoraphobia is so severe that he has not left his home for several years, even though he recognizes that he has missed out on several important family functions, like his granddaughter's graduation, and has given up his promising corporate career. Fred won't be moved to leave the house for everyday kinds of reasons, like a desire to meet a friend for a coffee at his old favorite spot. Fred may well recognize that there are plenty of reasons to justify leaving home on a daily basis, like having a coffee with a friend, but given his agoraphobia, he cannot act on these recognized reasons. There are some circumstances in which he may indeed leave his home—a raging fire or a convergence of visitors to his home that constitute a crowd. In those kinds of circumstances, Fred may not only judge that he has good reason to leave the house, but he may indeed face his fear and leave for that very reason.[12]

One way to capture the core of this disorder of agency in control terms is to note that such patients' threshold of reasons-reactivity is too high. They can recognize but cannot react appropriately to a wide variety of justifying reasons to do other than their disorder disposes them. It takes a rather strong justifying reason to get an agoraphobic patient to, say, leave her home (e.g., a raging fire or unwelcome house party), compared to the reasons for which non-clinical agents respond. As a contrast class, non-patient populations feature agents who not only recognize but can also act on a wide range of everyday incentives—what I'll term *weak justifying reasons* for action. For instance, if the action at issue is leaving one's home, these agents can do so for reasons such as meeting a friend for coffee, attending a talk in the city, or taking a jog through the park. What seems characteristic of the patient's exercise of agency, then, is this inability to respond to (sometimes

---

[12] The Fred case, as presented here, is a paraphrased version of the Fred case in Mele (2000), which has been expanded and modified here to fit the discussion of agoraphobia and control in the current context.

recognized) weak justifying reasons for action. The inability is what seems to be distinctive of the agoraphobic's agency and explains her restricted exercise of agency in both moral and nonmoral contexts.

This initial application of reasons-responsiveness terminology to the agoraphobic patient's action control requires further careful unpacking. The profile of the control capacities of the agent with agoraphobia above appeal to a distinction in justifying strength between strong and weak justifying reasons. Though both weak justifying reasons and strong justifying reasons are justificatory, there are many circumstances in which there are competing reasons for performing incompatible actions. To define strength of justifying reasons, we can look at the extent to which a reason R justifies an action A relative to a set of circumstances C. Hence, a *strong* justifying reason to A is a justifying reason to A (i.e., is sufficient for A) across a wide variety of circumstances. In contrast, a *weak* justifying reason to A is a justifying reason to A across many fewer circumstances. So a weak justifying reason to A is more likely to be defeated by competing reasons for performing incompatible actions (or at least for not A-ing) than a strong justifying reason to A is. For example, the desire to bake cupcakes may justify remaining at home in some circumstances. But if your home is engulfed in flames, your desire to remain home in order to bake cupcakes will give way to your desire not to perish in a fire. Obviously, a desire to bake cupcakes is a weak justifying reason in comparison to the desire not to perish in a fire. There are very few, if any, competing reasons that would justify staying in a burning building, provided one desired not to perish in a fire. Hence, the strength of weak and strong justifying reasons will vary in degree, roughly, in proportion to the set of circumstances in which they justify action.[13]

Of course, we needn't prescribe a monolithic view to explain the deficits of agency of all individuals with agoraphobia. Indeed, it is consistent with a reasons-responsive picture to accept that some persons with agoraphobia have deficits of reasons-receptivity. For instance, one might propose that when such an individual, pre-therapy, contemplates visiting or occupying a place in which escape or getting help is unlikely (e.g., a crowded subway car, a wedding party in a packed banquet hall), the anticipatory anxiety of doing

---

[13] In Waller (2014) and here my approach to reasons is akin to Sinnott-Armstrong's (2013) contrastivist account of practical reasons—that practical reasons to A are to be understood relative to a contrast class or set of alternatives. Moreover, a metaphysics of reasons according to which desires are justifying reasons is Humean in character. The above arguments, however, are compatible with alternative views about reasons.

so weakens the justificatory force of her desire to attend the (crowded) event.[14] So, although later—perhaps in conversation with her therapist—she can retroactively identify the desire to attend the event as a reason that justifies leaving home that day, it may be that in the moment she could not have recognized that desire as a reason to leave. Here the agent in the moment of action is not receptive to her reasons to leave her home to which she would otherwise be receptive. That is, in cases like this, the anxiety masks or modifies the agent's reasons-receptivity with respect to reasons to leave.

In either case, the agency of patients with agoraphobia can be characterized by deficits in control in terms of their responsiveness to reasons. And, insofar as attending to social and professional obligations are actions for which agents are typically held morally accountable, these reasons-responsiveness deficits can help explain why we may excuse or exempt an individual with agoraphobia from blame for missing these life events.

One might object to this reasons-responsiveness understanding of agoraphobia on the grounds that individuals with agoraphobia do not exercise control in the sense of choice and practical reasoning. Using the same case as an example, Ayob (2016) argues that, on the contrary, we can explain the agoraphobic agent's act of leaving his home in the event of a fire by appealing to the (presumably motivational) dominance of one competing urge over another—in this case urges associated with the fear of perishing in a fire and the anxiety of leaving home. Hence, Ayob (2016) argues that no choice was made to leave the house in the event of the fire. The debate here turns on which kind of mechanism, practical reasoning or otherwise, produces action in the case and further how to understand the notion of choice.

In reply, I see no reason to reject that one cannot choose in the sense of decide to leave a burning building in the case at hand, especially if one is subject to uncertainty about what to do (i.e., conflicting fears). Here we can helpfully distinguish decisions from *free* decisions, and note that the agent with agoraphobia in question made a choice, but not a free one. Individuals with agoraphobia might choose what to do in these scenarios without it being the case that they have chosen freely, a more robust control over their decisions and actions. Indeed the reasons-responsiveness approach captures this point well: they exercise control—are responsive to reasons—to a degree, but fall short of the threshold of control for responsible agency. In

[14] I am grateful to Joshua May and Matt King for their suggestion of this application of reasons-receptivity to the case.

contrast, a view according to which such individuals do not and cannot choose is a view that denies agency simpliciter to those individuals—at least which respect to their disorder-driven actions. As such, to deny agency is to deny even the intentional agency that Pickard attributes in cases of personality disorder. Support for this agency—however diminished—view of agoraphobic behavior can be further found in the clinical literature in which the avoidance behavior of individuals with agoraphobia is, on one model, understood as a kind of "goal-directed defensive action" and, moreover, where treatment of this avoidance is understood as targeting aspects of the patient beyond the conditioning of the learned association (Pittig et al. 2020).

We have characterized the agency of individuals with agoraphobia pre-therapy in terms of responsiveness to reasons. How does control as reasons-responsiveness change over the course of psychotherapy? Here we will focus on the first explanation of agoraphobia as involving too high of a threshold for reactivity to reasons. Individuals with agoraphobia require, on this view, a strong justifying reason to leave their home compared to the reasons for which nonclinical populations flexibly come and go between home and public places. Hence, if this is right, then the effective methods of psycho-therapy should act on those control capacities to react to everyday justifying reasons to be in public. And indeed, if we examine the most efficacious therapies for agoraphobia (and other anxiety disorders), we see that the effective techniques are exactly those that strengthen the patients' reasons-reactivity to weak justifying reasons to do other than their disorder directs. For instance, in Exposure Therapy (Roach and Foa 2006), patients are subject to graded exposure to fearful stimuli in either imagined scenarios, virtual reality, or in vivo—i.e., real scenarios (this is often paired with emotional desensitization techniques such as muscle relaxation exercises.).[15] For instance, the patient might be immersed in the real or imagined task of conversing with a small group of people outside of the home. Exposure to triggering stimuli is graded such that patients build up to confronting full-blown triggers.[16] Hence, the goal is to break reinforcement of learned avoidance responses so that patients can, post therapy, act on

---

[15] For a recent review of the efficacy of Exposure Therapy for agoraphobia, see Buchholtz and Abramowitz (2020). For an updated review of the use of virtual reality in therapy, see Kim and Kim (2020).

[16] In this way, graded exposure as a technique differs from Implosive (flooding) Therapy in which patients are exposed to the triggering stimuli in its entirety even in initial sessions of therapy (see, e.g., Morganstern 1973).

recognized everyday reasons to be near once fearful stimuli—to be reactive to common (i.e., weak) justifying reasons for action.

Now let's return to cases of agoraphobia in which anticipatory anxiety of crowds or confined space blocks or modifies the agent's *recognition of reasons* that justify leaving home. In these cases the deficit of agency seems to concern reasons-receptivity. The agent with agoraphobia has difficulty recognizing everyday reasons for leaving home due to the experience of anxiety. Here, too, several techniques of Exposure Therapy target and modify the patients' reasons-receptivity. For instance, at the heart of Exposure Therapy is repeated and graduated exposure to the feared stimulus (imagined, virtual, or actual). One proposed mechanism of symptom reduction is habituation of fear: anxiety of the feared stimulus is activated during exposure in sessions, and over the course of therapy fear is attenuated (Foa and Kozak 1986). Another proposed mechanism of symptom reduction is "expectation violation." According to this theory, during exposure to the feared stimulus in therapy what the patient expects to occur (e.g., the panic-like symptoms) does not occur. Since the adverse outcome is absent, the patient's avoidance of the stimuli (e.g., crowded places) is lessened (Craske et al. 2014). Further, we can look to other techniques paired with exposure in Exposure Therapy as playing a role in dampening the agent's anxiety. Exposure to the feared stimulus is often paired with training for opposite responses, such as muscle relaxation techniques. One proposal is that the opposite pairing of a relaxation technique with triggered anxiety during exposure inhibits the anxious response (Wolpe 1968). Hence, regardless of the explanatory mechanism, on these accounts the exposure or relaxation techniques directly or indirectly change the individual's anxiety about the feared stimuli. When anxiety about the stimulus is attenuated, patients are more easily able to recognize reasons, such as a desire to have coffee with a friend or a desire to attend a family wedding, as justifying leaving one's home in the moment.

4.6 Personality Disorder and Dialectical Behavior Therapy

The resources of reasons-responsive accounts of control can be extended to understand the deficits of agency and so responsibility of patients with other psychopathologies, such as addiction, obsessive-compulsive disorder, and personality disorder. Here I'll briefly make the case for a richer account of the control deficits of patients with personality disorder and of why personality disorder is successfully addressed with Dialectical Behavior Therapy, another third-wave cognitive-behavioral therapy, in particular.[17]

First, we must gain a clear understanding of the complex category of personality disorder. It's helpful here to refer to work of philosophers who are also clinical practitioners. Nancy Potter, for instance, describes personality disorder as a "structural organization of the personality that is exhibited by a dysfunctional pattern of behavior" (2009, 2). By personality here, clinicians mean the characteristic manner in which a person thinks, feels, behaves, and relates to others. This pattern of behavior is characteristically maladaptive, stable, and inflexible. Hannah Pickard writes that, "PD [personality disorder] occurs when the set of characteristics or traits that make a person the kind of person that they are causes severe psychological distress and impairment in social, occupational, or other important contexts" (2011, 181). Some associated behaviors include risk-taking, substance abuse, inability to maintain interpersonal relationships, impulsivity (which may lead to switching jobs or inability to hold down a job), harmful and manipulative behavior, and self-harm.

Here I'll narrow my focus to borderline personality disorder in particular, distinguishable in terms of diagnosis from (although frequently co-morbid with) psychopathy and other personality disorder diagnoses. The DSM-5 stipulates the following criteria for diagnosis of borderline personality disorder:

a) A pervasive pattern of instability of interpersonal relationships, self-image, and affects, and marked impulsivity, beginning by early adulthood and present in a variety of contexts, as indicated by five (or more) of the following:

b) Frantic efforts to avoid real or imagined abandonment.

c) A pattern of unstable and intense interpersonal relationships characterized by alternating between extremes of idealization and devaluation.

d) Identity disturbance: markedly and persistently unstable self-image or sense of self.

e) Impulsivity in at least two areas that are potentially self-damaging (e.g., spending, sex, substance abuse, reckless driving, binge eating).

---

[17] See Horne (2014) for a discussion of Dialectical Behavior Therapy as a window to understanding the *moral* deficit model of personality disorder.

f) Recurrent suicidal behavior, gestures, or threats, or self-mutilating behavior.

g) Affective instability due to a marked reactivity of mood (e.g., intense episodic dysphoria, irritability, or anxiety usually lasting a few hours and only rarely more than a few days).

h) Chronic feelings of emptiness.

i) Inappropriate, intense anger or difficulty controlling anger (e.g., frequent displays of temper, constant anger, recurrent physical fights).

j) Transient, stress-related paranoid ideation or severe dissociative symptoms. (DSM-5, 2013, 663, abbreviated excerpt)

An article on the what-its-like-ness of borderline personality disorder notes it is sometimes called "chronic irrationality... Think severe mood swings, impulsivity, instability, and a whole lot of explosive anger" (Marlborough 2016, Vice). The impulsivity of borderline personality disorder isn't restricted to impulsive and angry outbursts but extends to unsafe sexual practices, reckless driving, spending sprees, frequent changes in employment, or binging. Individuals with borderline personality disorder face an inner conflict concerning self-image, marked by self-devaluation while yet striving for normality (Spodenkiewicz et al. 2013), and this inner conflict can lead to unstable relationships marked by extremes of attachment. In psychiatric contexts, this borderline personality disorder symptom profile is often explained in part in terms of the constructs of *low distress tolerance* and *high experiential avoidance*.[18] That is, patients with borderline personality disorder exhibit an abnormal unwillingness to remain in contact with aversive private experiences (e.g., certain bodily sensation, thoughts, emotions, memories) and so are disposed to engage in behavior, such as these impulsive ones, that alters those aversive experiences. (These constructs are two of the maladaptive emotion-regulation strategies mentioned in Section 4.2.)

Drawing upon reasons-responsive accounts of control, we can isolate the control deficits of patients with borderline personality disorder as exhibiting

---

[18] See, e.g., Chapman, Specht, and Celluci (2005). See Bolton and Banner (2012) for a related discussion of experienced distress as a factor in the impaired agency of those with mental disorders. For other psychiatric constructs that have been proposed as explanatory in the case of personality disorder, see Lineman (1993) on emotion dysregulation and Fonagy and Bateman (2006) on lack of mentalization.

too low of a threshold for reasons-reactivity with respect to distress. This is particularly the case in the interpersonal domain. Patients with borderline personality disorder are more likely to take everyday distress to justify engaging in certain harmful and impulsive actions compared to non-patient populations. There are only a few efficacious talk therapies for borderline personality disorder, but one is Dialectical Behavior Therapy. Dialectical Behavior Therapy involves the introduction and honing of skills for mindfulness, emotion regulation, distress tolerance, and interpersonal effectiveness (Linehan 1993). For example, a therapist may teach patients how to tolerate pain in difficult situations and how to change their emotions that they desire to change in those circumstances. Hence, instead of taking distress as a justifying reason to pursue risky behavior or to give up on meaningful relationships and jobs, patients learn to tolerate or change certain distress reactions and so raise the bar of what justifies action. For example, patients in Dialectical Behavior Therapy practice radical acceptance. Instead of thinking on how one would like to change an undesirable situation that is out of one's control (e.g., job loss, relationship split), patients focus instead on acknowledging their frustration while also accepting that it is so. This acceptance is facilitated by engaging in an alternate activity to redirect or distract, such as comparing the situation to a worse contrast case or, alternatively, pursuing activities that keep one otherwise focused or change one's mood in a positive direction. If efficacious, Dialectical Behavior Therapy patients should see a reduction in their experiential avoidance and hence healthier agency, both intrapersonally and interpersonally, in their daily lives. They will, for example, engage in less risky and impulsive actions and maintain more stable employment and relationships. This ability to maintain close relationships will facilitate their participation in the moral responsibility practices (e.g., healthier familial, peer, and romantic relationships).

Patients in Dialectical Behavior Therapy, who previous to therapy may have acted on distress as a justifying reason, learn how to avoid acting maladaptively in the face of distress, even if distress remains present. In reasons-responsiveness parlance, through therapy patients work to adjust their reasons-reactivity capacities. Moreover, the Dialectical Behavioral Therapy patient's reasons-receptivity may change as well; insofar as patients can in instances dampen their distress beyond radical acceptance, those patients may no longer consider distress to be a justifying reason to act. If so, agents with borderline personality disorder post-therapy may more closely approximate the distress tolerance and so reasons-responsiveness profile of nonclinical populations.[19]

Of note, this framework for understanding therapeutic change for borderline personality disorder as involving changes in reasons-responsiveness comports with Horne's (2014) suggestion that borderline personality disorder patients, pre-therapy, can be aware of moral reasons for action, even if they cannot react to those recognized reasons.[20] We can especially see that this is the case if we return to the impairments in self and self-direction that are characteristic of the patient with borderline personality experience. Such patients strive for normalcy and can see the harm caused by their actions; yet they continue to act in these destructive ways, which can lead to psychic conflict. This is indicative of an agent who can recognize moral reasons for action but cannot act on them.

That agents with borderline personality disorder recognize moral reasons is consistent with various possible relations to those reasons. One possibility is that such individuals recognize that harm to others is a reason not to act in relationship-destroying ways, yet those reasons lack the motivational weight in the moment of decision that they typically hold for other agents. Another possibility is that the moral reasons possess a motivational weight for the individual in light of her goals for her self-expression, but that irrationally she acts on weaker motivational reasons. A third possibility concerns the fact that, phenomenally, those moral reasons at times seem to feature in rumination and guilt of the agent with borderline personality exclusively, particularly for impulsive actions. As such, the recognition of moral reasons might be only a dispositional feature of the agent pre-action but becomes occurrent or conscious in the rumination and guilt after the individual acts destructively. In any case, the fact that patients with borderline personality disorder experience profound distress and guilt over their impulsive and destructive actions and that their endorsed values are for acting otherwise is highly suggestive of an agent who recognizes in a robust sense moral reasons for action, even if they do not always act in line with those reasons. As Marlborough (2016) notes in his first-person account of borderline

---

[19] Indeed, one might suspect that the anxiety and avoidance of a patient with agoraphobia could be understood similarly in terms of distress tolerance and experiential avoidance. Studies of these constructs in a wide-range of clinical populations with anxiety support that anxiety sensitivity and these two constructs overlap but are not synonymous (Keough et al. 2010; Kämpfe et al. 2012).

[20] "Lineman's experience suggests that BPD patients are fully aware, and even knowledgeable, about moral ways to behave with respect to one another . . . Their problem is application of this knowledge" (Horne 2014, 13).

personality disorder: "Of course, the outburst didn't give me any sense of relief. It turned into a looping internal monologue of personal recrimination and self-hatred. Every decision is retroactively punished." Here, then, we can understand the inner reflection and behavioral patterns of an agent with borderline personality disorder as reflecting deficits in reasons-reactivity which then manifest in disordered agency.

4.7 Concluding Remarks

I have argued that the agential deficits of select clinical populations, who may be excused or partially exempt from our moral practices, can be illuminated via a control-centric framework. I have outlined that framework by providing an understanding of disorder-specific reasons-responsive profile, both pre-talk therapy and post-talk therapy. We can see effective techniques of talk therapy, such as gradual exposure or radical acceptance exercises, as operating on the reasons-responsive capacities of patients. In this way, reasons-responsiveness is a lens through which both to better appreciate the flourishing via control of agents in both moral and nonmoral contexts and to understand psychiatric explanatory constructs.

Moreover, there are implications for our moral community. As has been argued here and by others (including Shoemaker 2015 and Horne 2014), agents who are members of clinical populations may meet certain conditions for being part of our moral community in rich and varied ways. This study of the change in reasons-responsible capacities over a course of talk therapy throws into relief the initial borderline or marginal status and transition into the circle of full-blown moral agency for individuals in those clinical groups. As demonstrated, the control deficits exhibited by patients with agoraphobia and borderline personality disorder can lead to interpersonal difficulties. A pronounced and sustained history of interpersonal difficulties can leave an agent at odds with other agents and so at the fringes of a moral community. That is, control deficits can lead to strained participation in moral life. Enhancing control capacities allows for fuller participation in interpersonal contexts such as navigating relations in a moral community.[21]

[21] I am grateful to Matt King and Joshua May for their generous and illuminating feedback on several drafts of this chapter. I'd also like to thank colleagues at departmental research seminars at Franklin and Marshall College and Iona College as well as audience members at the 2018 Society for Philosophy and Psychology and the 2019 Joint Session of the Aristotelian Society and the Mind Association for their helpful comments.

References

Aldao, A., & Nolen-Hoeksema, S. (2012). When are adaptive strategies most predictive of psychopathology?. *Journal of Abnormal Psychology 121*(1), 276.

Aldao, A., Nolen-Hoeksema, S., & Schweizer, S. (2010). Emotion-regulation strategies across psychopathology: A meta-analytic review. *Clinical Psychology Review 30*(2), 217–237.

American Psychiatric Association. (2013). *Diagnostic and Statistical Manual of Mental Disorders (DSM-5®)*. Washington, DC: American Psychiatric Publishing.

Ayob, G. (2016). Agency in the absence of reason-responsiveness: The case of dispositional impulsivity in personality disorders. *Philosophy, Psychiatry, and Psychology 23*(1), 61–73.

Beck, A. (1976). *Cognitive Therapy and the Emotional Disorders*. New York: International University Press.

Bolton, D., & Banner, N. (2012). Does mental disorder involve loss of personal autonomy. In Radoilska, L. (ed.) *Autonomy and Mental Disorder*, 77–99. Oxford: Oxford University Press.

Buchholtz, J. L., & Abramowitz, J. S. (2020). The therapeutic alliance in exposure therapy for anxiety-related disorders: A critical review. *Journal of Anxiety Disorder 70*.

Chapman, A.L., Specht, M.W., & Cellucci, T. (2005). Borderline personality disorder and deliberate self-harm: Does experiential avoidance play a role?. *Suicide and Life-Threatening Behavior 35*(4), 388–399.

Charland, L. C. (2004). Moral treatment and the personality disorders. In J. Radden (ed.), *The Philosophy of Psychiatry: A Companion*, 64–78. Oxford: Oxford University Press.

Craske, M. G., Treanor, M., Conway, C. C., Zbozinek, T., & Vervliet, B. (2014). Maximizing exposure therapy: An inhibitory learning approach. *Behaviour Research and Therapy 58*, 10–23.

Ellis, A. (1969). A cognitive approach to behavior therapy. *International Journal of Psychiatry 8*, 896–900.

Fischer, J. M. (2006). *My Way: Essays on Moral Responsibility*. Oxford: Oxford University Press.

Fischer, J. M., & Ravizza, M. (1998). *Responsibility and Control: A Theory of Moral Responsibility*. Cambridge: Cambridge University Press.

Foa, E. B., & Kozak, M. J. (1986). Emotional processing of fear: Exposure to corrective information. *Psychological Bulletin 99*(1), 20.

Fonagy, P., & Bateman, A. (2006). Progress in the treatment of borderline personality disorder. *The British Journal of Psychiatry 188*(1), 1–3.

Frankfurt, H. G. (1971). Freedom of the will and the concept of a person. *Journal of Philosophy 68 (1),* 5–20.

Gallagher, M. W., Naragon-Gainey, K., & Brown, T. A. (2014). Perceived control is a transdiagnostic predictor of cognitive–behavior therapy outcome for anxiety disorders. *Cognitive Therapy and Research 38*(1), 10–22.

Gillanders, D. T., Bolderston, H., Bond, F. W., Dempster, M., Flaxman, P. E., Campbell, L., . . . Remington, B. (2014). The development and initial validation of The Cognitive Fusion Questionnaire. *Behavior Therapy 45*, 83–101.

Glannon, W. (2017). Psychopathy and responsibility: Empirical data and normative judgments. *Philosophy, Psychiatry, and Psychology 24 (1),* 13–15.

Godman, M. & Jefferson, A. (2017). On blaming and punishing psychopaths. *Criminal Law and Philosophy 11*(1), 127–142.

Gorman, A. (2020). Depression's Threat to Self-Governance. *Social Theory and Practice 46*(2), 277–297.

Greenspan, P. S. (2003). Responsible psychopaths. *Philosophical Psychology 16* (3), 417–429.

Gross, J. J., & John, O. P. (2003). Individual differences in two emotion regulation processes: Implications for affect, relationships, and well-being. *Journal of Personality and Social Psychology 85*(2), 348.

Haji, I. (1998). *Moral Appraisability*. New York: Oxford University Press.

Hayes, S. C. (2004). Acceptance and commitment therapy, relational frame theory, and the third wave of behavioral and cognitive therapies. *Behavior Therapy 4*(35), 639–665.

Hayes, S. C., Strosahl, K., D., & Wilson, K. G. (2011). *Acceptance and Commitment Therapy (2nd Edition): The Process and Practice of Mindful Change.* New York: Guilford Press.

Hofmann, S. G., & Asmundson, G. J. (2008). Acceptance and mindfulness-based therapy: New wave or old hat?. *Clinical Psychology Review 28*(1), 1–16.

Horne, G. (2014). Is Borderline Personality Disorder a moral or clinical condition? Assessing Charland's argument from treatment. *Neuroethics 7*(2), 215–226.

Jefferson, A. & Sifferd, K. (2018). Are psychopaths legally insane? *European Journal of Analytic Philosophy 14*(1), 79–96.

Kämpfe, C. K., Gloster, A. T., Wittchen, H. U., Helbig-Lang, S., Lang, T., Gerlach, A. L., . . . & Hamm, A. O. (2012). Experiential avoidance and anxiety sensitivity in patients with panic disorder and agoraphobia: Do both constructs measure the same?. *International Journal of Clinical and Health Psychology 12*(1), 5–22.

Keogh, E., Book, K., Thomas, J., Giddins, G., & Eccleston, C. (2010). Predicting pain and disability in patients with hand fractures: Comparing pain anxiety, anxiety sensitivity and pain catastrophizing. *European Journal of Pain 14*(4), 446–451.

Kiehl, K. A. & Sinnott-Armstrong, W. P. (eds.) (2013). *Handbook on Psychopathy and Law*. Oxford: Oxford University Press.

Kim, S., & Kim, E. (2020). The use of virtual reality in psychiatry: A review. *Journal of the Korean Academy of Child and Adolescent Psychiatry 31*(1), 26–32.

King, M., & May, J. (2018). Moral responsibility and mental illness: A call for nuance. *Neuroethics 11*(1), 11–22.

Levy, N. (ed.). (2013). Addiction and self-control: Perspectives from philosophy, psychology, and neuroscience. Oxford: Oxford University Press.

Levy, N. (2014). Psychopaths and blame: The argument from content. *Philosophical Psychology 27*(3), 351–367.

Linehan, M. M. (1993). *Skills Training Manual for Treating Borderline Personality Disorder*. New York: Guilford Press.

Marlborough, Patrick. (2016). What it's like to have Borderline Personality Disorder. *Vice*. URL: https://www.vice.com/en_us/article/9b83wa/how-it-feels-to-suffer-borderline-personality-disorder.

McKenna, M. (2012). *Conversation and Responsibility*. Oxford: Oxford University Press.

Mele, A. R. (2000). Reactive attitudes, reactivity, and omissions. *Philosophy and Phenomenological Research 61*(2), 447–452.

Morganstern, K. P. (1973). Implosive therapy and flooding procedures: A critical review. *Psychological Bulletin 79*(5), 318.

Nadelhoffer, T. & Sinnott-Armstrong, W. (2013). Is psychopathy a mental disease? In Nicole Vincent (ed.), *Neuroscience and Legal Responsibility*. Oxford: Oxford University Press, pp. 229–255.

Nelkin, D. K. (2011). *Making Sense of Freedom and Responsibility*. Oxford: Oxford University Press.

Nelkin, D. K. (2017). Fine cuts of moral agency: Dissociable deficits in psychopathy and autism. In S. Matthew Liao & Collin O'Neil (eds.), *Current Controversies in Bioethics*. New York: Routledge, pp. 47–66.

Pearce, S., & Pickard, H. (2009). The moral content of psychiatric treatment. *The British Journal of Psychiatry 195*(4), 281–282.

Pickard, H. (2011). Responsibility without blame: Empathy and the effective treatment of personality disorder. *Philosophy, Psychiatry, & Psychology: PPP* 18(3), 209.

Pickard, H. (2013). Responsibility without blame: Philosophical reflections on clinical practice. *Oxford Handbook of Philosophy of Psychiatry.* Oxford: Oxford University Press, 1134–1154.

Pickard, H. (2015). Psychopathology and the ability to do otherwise. *Philosophy and Phenomenological Research* 90(1), 135–163.

Pickard, H. (2017). Responsibility without blame for addiction. *Neuroethics 10* (1), 169–180.

Pittig, A., Wong, A. H., Glück, V. M., & Boschet, J. M. (2020). Avoidance and its bi-directional relationship with conditioned fear: Mechanisms, moderators, and clinical implications. *Behaviour Research and Therapy 126*, 103, 550.

Potter, N. N. (2009). *Mapping the Edges and the In-Between: A Critical Analysis of Borderline Personality Disorder.* Oxford: Oxford University Press.

Rauch S., & Foa, E. B. (2006). Emotional processing theory (EPT) and exposure therapy for PTSD. *Journal of Contemporary Psychotherapy 36*(2), 61–65.

Sartorio, C. (2016). *Causation and Free Will.* Oxford: Oxford University Press.

Segal, Z. V., Williams, J. M. G., & Teasdale, J. D. (2002). *Mindfulness-Based Cognitive Therapy for Depression.* New York: Guilford Press.

Shoemaker, D. (2015). *Responsibility from the Margins.* Oxford University Press, USA.

Sinnott-Armstrong, W. (2013). Free contrastivism. In Blaauw, M. (ed.) *Contrastivism in Philosophy.* New York: Routledge.

Sloan, E., Hall, K., Moulding, R., Bryce, S., Mildred, H., & Staiger, P. K. (2017). Emotion regulation as a transdiagnostic treatment construct across anxiety, depression, substance, eating and borderline personality disorders: A systematic review. *Clinical Psychology Review 57*, 141–163.

Spodenkiewicz, M., Speranza, M., Tieb, O., Pham-Scottez, A., Corcos, M., & Revah-Levy, A. (2013). Living from day to day—qualitative study on borderline personality disorder in adolescence. *Journal of the Canadian Academy of Child and Adolescent Psychiatry 22*(4), 282–289.

Strawson, P. F. (1974). *Freedom and Resentment, and Other Essays* (Vol. 595). London: Egmont Books (UK).

Summers, J. S., & Sinnott-Armstrong, W. (2019). *Clean Hands: Philosophical Lessons from Scrupulosity.* Oxford: Oxford University Press.

Wallace, R. J. (1994). *Responsibility and the Moral Sentiments.* Cambridge, MA: Harvard University Press.

Waller, R. R. (2014). Revising reasons reactivity: Weakly and strongly sufficient reasons for acting. *Ethical Theory and Moral Practice 17*(3), 529–543.

Watson, G. (1975). Free agency. *Journal of Philosophy 72*, 205–20.

Watson, G. (2004). *Agency and Answerability: Selected Essays.* Oxford: Oxford University Press.

Watson, G. (2013). XIV—Psychopathic agency and prudential deficits. *Proceedings of the Aristotelian Society 113* (3pt3), 269–292.

Wolf, S. (1987). Sanity and the metaphysics of responsibility. In Schoeman, F. (ed.). *Responsibility, Character, and the Emotions: New Essays in Moral Psychology.* Cambridge: Cambridge University Press.

Wolf, S. (1990). *Freedom within Reason.* Oxford: Oxford University Press.

Wolpe, J. (1968). Psychotherapy by reciprocal inhibition. *Conditional Reflex: A Pavlovian Journal of Research & Therapy 3*(4), 234–240.

Wonderly, M. (forthcoming). Psychopathy, agency, and practical reason. In Ruth Chang & Kurt Sylvan (eds.), *Routledge Handbook of Practical Reason.* New York, USA: Routledge.

5

Legal Insanity and Moral Knowledge

Why Is a Lack of Moral Knowledge Related to a Mental Illness Exculpatory?

Katrina L. Sifferd

In this chapter I argue that a successful plea of legal insanity ought to rest upon proof that a criminal act is causally related to symptoms of a mental disorder. Although mental disorder diagnoses and legal insanity are concepts crafted to serve different purposes, and thus often may refer to different mental incapacities, the former signal causal-historical information relevant to the latter. The canonical case of legal insanity involves a defendant who forms the requisite *mens rea* for a crime but lacks understanding of the legal and moral quality of her act—typically, that all things considered the act is wrong (Morse and Bonnie 2013). Symptoms of a mental disorder can undermine a person's capacity to be law-abiding at the time of a crime by causing a lack of moral knowledge; and the presence of a mental disorder signals to the court that the defendant is not culpable for this ignorance. Other cases of moral ignorance or incompetence indicate that a lack of moral knowledge can also be due to miseducation or other extreme environmental conditions unrelated to a mental disorder (Wolf 1987). The child of a ruthless dictator, for example, might seem "insane" due to this lack of moral knowledge. I will argue, however, such cases ought not to underpin a claim of legal insanity. To be law-abiding, persons with the capacity to be legally responsible (Hart 1968) are required to understand the criminal law's moral demands, to reflect on these demands across time, and to control one's behavior in light of them. When people fail to do so, they are culpable for their ignorance.

Katrina L. Sifferd, *Legal Insanity and Moral Knowledge: Why Is a Lack of Moral Knowledge Related to a Mental Illness Exculpatory?* In: *Agency in Mental Disorder: Philosophical Dimensions.* Edited by: Matt King & Joshua May, Oxford University Press. © Katrina L. Sifferd 2022. DOI: 10.1093/oso/9780198868811.003.0006

5.1 The Gap between Mental Disorder Diagnoses and Legal Insanity

Although I believe that mental health diagnoses are relevant to criminal responsibility, it is important to recognize that the legal concept of insanity and diagnoses made under the psycho-medical umbrella concept "mental disorder" serve very different purposes. For this reason we ought not to expect mental disorder diagnoses (even diagnoses of schizophrenia or psychosis) and the legal category of insanity to refer to the same mental incapacities or deficits (Jefferson and Sifferd 2018; Moore 2015; Morse 2015; Hirstein, Sifferd, and Fagan 2018). Instead, certain diagnoses can only indicate that a defendant may have suffered from symptoms at the time of the crime that could have resulted in that defendant being legally insane. Whether those symptoms resulted in the defendant being legally insane with regard to the criminal act requires fact-finding by the court.

5.1.1 The Aim of Mental Disorder Diagnoses

Psychiatrists and psychologists craft mental disorder categories for the purpose of diagnosing and treating mental health problems against a social background of what is considered a normal or healthy life. It is common practice to articulate the requirements for diagnoses with the goal of clustering symptoms related to a common underlying etiology or cause; but it is clear from the history of psychological diagnoses that mental disorders often identify a heterogeneous class of symptoms and etiologies that can later fragment in new diagnoses. For example, what is now the diagnosis of autism was initially referred to as childhood psychosis, and then as a severe mental developmental disorder. Autism now refers to a spectrum disorder ranging from mild to very severe levels of impairment usually involving social difficulties. The diagnosis of psychopathy, which is noted as a "specifier" of anti-social personality disorder in the Diagnostic and Statistical Manual of Mental Disorders V, is an example of a diagnosis that seems to fail in identifying a common cause or even a common cluster of symptoms. In the psychological literature the term seems to identify a very heterogenous class of persons even with regard to the central symptoms of flattened affect or lack of empathy (Jefferson and Sifferd 2018). When persons deemed psychopaths by the primary diagnostic (the Psychopathy Checklist-Revised) are split into subgroups such as "successful" and "unsuccessful," or "primary" and "secondary," their symptoms look very different.

In general, one can find persons with very different clusters of symptoms within most categories of mental disorder. One patient with autism may be non-verbal and find social interaction extremely difficult, whereas another might have only mild difficulty with social interactions but have difficulty task-switching and with other types of cognitive flexibility. As indicated above, this may be due to a lack of a common etiology; but it may also be because many mental disorders (including autism) operate on a continuum with multiple symptoms each ranging from mild to very severe (King and May 2018). In addition, disorders can be episodic, meaning their symptoms can come and go: they present themselves in (more or less) discrete instances and thus can be environmentally contingent (King and May 2018). Even assuming a common etiology, a child with autism may exhibit very different symptoms in a loud, chaotic family home compared to a very structured school designed for children with autism. Other examples of episodic disorders might include post-traumatic stress disorder (PTSD), bipolar disorder, and phobias. However, some mental disorders are more static, possibly including depression and generalized anxiety disorder. For these disorders, manifestation of symptoms "is more likely to persist over time" but with no clear boundaries (King and May 2018).

A person's particular set of symptoms thus depends not only on their diagnosis, but also often on each symptom's location on a continuum of severity; whether their mental disorder is episodic or static; and the persons' environment(s). Thus, persons with the same diagnosis can have different levels of capacity and incapacity at different times, in different environments, and with regard to different domains of action. A person with obsessive-compulsive disorder (OCD) or agoraphobia may find it very difficult to leave their home; but be very capable of working as an accountant, or computer programmer, from home. Two persons with OCD can have very different levels of impairment, depending on whether they have undergone treatment, are taking medication, and are within an environment that exacerbates or alleviates their symptoms. One person with anxiety may only suffer incapacities when experiencing an anxiety attack; but have little impairment at other times and in other situations; others may go through periods where their anxiety is pervasive over time and across environments.

As stated above, the purpose of a mental disorder diagnosis is usually to identify and cluster symptoms so as to design a treatment or management regimen to address a patient's particular impairments related to a mental disorder given the environments within which they usually live and work. For all of these reasons, a mental disorder diagnosis does not carry enough information about the capacities relevant to legal culpability for the court to determine whether a defendant is legally insane.

5.1.2   The Aim of the Legal Insanity Defense

In contrast to the goal of diagnosis and treatment, the concept of legal insanity is closely aligned with the purposes of criminal law and punishment. The aim of the criminal law is to punish wrongdoing, reduce crime, or—on a "hybrid" theory—both. In the US, the language of the Model Penal Code (MPC) indicates retribution is the primary aim of the criminal law, and deterrence is a secondary aim (April 9, 2007).[1] In general, justifications and excuses to legal responsibility, including legal insanity, are intended to pick out cases where imposition of criminal punishment would not serve the purposes of retribution and deterrence. Defendants eligible for a justification or excuse are disqualified from criminal punishment either because their acts are not morally wrongful (in some cases, due to features of the actor, in other cases, due to features of the act) and/or because punishing such persons will not reduce crime.

A defendant may be found guilty of a crime only if both the *actus reus* (act) and *mens rea* (mental state) requirements are met. For example, under the MPC a person may only be found guilty of murder if he commits an act that (1) causes the death of another (2) for the purpose of causing that death or knowing that death would result. The presence of the mental state elements of the crime in addition to a criminal act is thus necessary for criminal responsibility. But, as justifications and excuses show, proof of *mens rea* and *actus reus* is not sufficient for guilt. For example, justifications negate wrongdoing. A person may kill another in self-defense, for instance, and thereby cause the death of another knowing that death would result, and yet they do nothing wrong and thus ought not be found guilty of a crime because the act is "justified." By contrast, affirmative defenses like legal insanity negate the responsibility of those whose conduct both was wrongful

---

[1] The Model Penal Code (MPC) is a text initially published in 1962 as a project by the American Law Institute with the aim of assisting U.S. state legislatures to update and standardize the penal law of the United States of America. Over two-thirds of U.S. states have used the MPC to revise and replace portions of their criminal codes.

and included the requisite *mens rea*. Another way of putting this is to say an affirmative defense claims a person is not culpable even if they intended to commit, and were successful in committing, a wrongful act.

Legal insanity is considered to be an excuse. Consider the famous M'Naghten case. Due to a mental illness, defendant M'Naghten held the false belief that the British Tory party planned to kill him, and that he needed to kill the Prime Minister to end the threat to his life (Morse and Bonnie 2013). M'Naghten thus did indeed intend to kill Prime Minister Peel, and acted in furtherance of this intention, even though his reasons for forming this intent were delusional (Morse and Bonnie 2013, 491). M'Naghten was found legally insane. Similarly, infamous US defendant Andrea Yates fully intended to kill her five children when she drowned them one by one; however, she believed she was saving them from eternal damnation by doing so (Morse and Bonnie 2013, 492). Yates was also found to be excused due to legal insanity.

Both M'Naghten and Yates held delusional beliefs caused by a mental illness resulting in a lack of moral knowledge about their crime. H.L.A. Hart proposes we understand criminal excuses like legal insanity by framing the criminal law as a *choosing system* (Hart 1968). According to Hart, a criminal excuse is best understood as "a mechanism for... maximizing within the framework of coercive criminal law the efficacy of the individual's informed and considered choice in determining the future and also his power to predict that future" (Hart 1968, 54). Recognition of excuses ensures that individuals will not be liable for criminal consequences that they did not choose because they lacked an understanding of the act due to a mental illness. Recognition of excusing conditions is therefore seen as a matter of protection of the individual, and this protection extends even to cases where the deterrent aim of criminal punishment may be served by holding a mentally ill person responsible (Hart 1968, 49).

Legal doctrine indicates that the presence of a mental disorder may be relevant to the culpability of certain defendants because the purposes of punishment—retribution, deterrence, incapacitation, and rehabilitation—are not met by punishing them. M'Naghten and Yates were unlikely to be deterred from their crime by threat of punishment because they did not understand their actions as wrong. However, the general population may be deterred from committing crimes by punishing the mentally ill, and inca-pacitation of persons who have committed crimes while they are mentally ill can decrease recidivism rates if a specific defendant would have committed more crimes if released. Since application of the legal insanity excuse is not conditional on the efficacy of deterrence, the best explanation of the excuse is that punishing insane offenders is fundamentally incompatible with retribution because the legally insane are not morally blameworthy or culpable.[2]

This understanding of the legal insanity defense coheres with Hart's understanding of excuses as protecting those who lack the ability to choose to be law abiding. The traditional common law test for legal insanity, called the M'Naghten rule after the M'Naghten's case, excuses a defendant who, due to a severe mental disease or defect, is unable to appreciate the nature and quality (namely, the wrongfulness) of his act.[3] Retribution requires that punishment be proportional to both the seriousness of the crime and the culpability of the offender. If the offender lacks important information about his act due to a mental illness—namely, knowledge that the act is wrong, or "moral knowledge" regarding the act—punishment may be inappropriate because it does not deliver a defendant's "just deserts."[4]

The insanity defense provides a criminal defendant an important opportunity to indicate his psychological states were compromised at the time he

---

[2] According to the MPC, retribution is the primary aim of criminal punishment.

[3] Common law is derived from judicial decisions or opinions, sometimes dating back to English court decisions. The M'Naghten rule comes from M'Naghten's Case, 8 Eng. Rep. 718, 722 (1843). The other test for legal insanity used in the U.S., found in the Model Penal Code (1985), is similar to the M'Naghten rule but has an additional volitional prong. This test for legal insanity requires that at the time of the criminal act a defendant diagnosed with a relevant mental defect lacked "substantial capacity to either appreciate the criminality of his conduct or to conform his conduct to the requirements of the law." The addition of a volitional prong has been controversial. Morse (1994) worries that it is difficult to formulate a test for when a person could not have done other than they did; in most cases people retain some measure of volitional control. My argument will focus on the M'Naghten test for legal insanity and the relationship between moral and legal knowledge and mental illness.

[4] Legal scholar Christopher Slobogin has argued that the legal insanity defense is unnecessary. A claim of legal insanity is not available to defendants who lack moral knowledge for reasons other than mental illness, including those who perform a criminal act in a state of extreme stress or a "blind rage" (Slobogin 2000). Defendants who lack moral knowledge, whatever the reason, have the opportunity to argue that they lack the mental state necessary for the crime (Slobogin 2000). In this case a successful negation of mens rea—due to lack of moral knowledge—would result in the defendant's acquittal.

One problem with this idea of using negation of mens rea as a substitute for legal insanity is that it excludes the canonical case of legal insanity from the defense. Both M'Naghten and Yates formed the requisite mens rea—they acted for the purpose of causing another's death. M'Naghten and Yates would have been convicted of murder despite their delusional beliefs and related failure to understand the moral quality of their acts had a plea of legal insanity not been available; so, at least some criminal defendants are insufficiently protected from unjust punishment by the demand that prosecutors prove mens rea and actus reus (Morse and Bonnie 2013). A prosecutor could meet the burden of proving the elements of a crime beyond a reasonable doubt, and yet the defendant's criminal responsibility could fail to be established because he lacked important (moral) information relevant to the act (Morse and Bonnie 2013).

committed a crime—even if he had the intent to cause criminal harm. In addition, the legal insanity plea allows the defendant to show *why* his choice-making was compromised—due to a mental disorder—in contrast with other reasons that are not generally considered exculpatory by the courts.

5.1.3 The Relationship between Mental Disorder and the M'Naghten Test for Legal Insanity

Under the M'Naghten test many persons with a mental disorder will still be fully responsible for criminal acts because at the time of their crime the disorder did not impact their moral knowledge. A successful plea of insanity typically requires a tight relationship between particular symptoms of a serious mental disorder, such as experiencing hallucinations or delusions, and the mental states causally related to the criminal act. Thus, in attempting to prove legal insanity, the defense offers proof of a mental disorder, and that symptoms related to this disorder resulted in failure of the defendant to understand the wrongfulness of the criminal act.

To better understand why a lack of moral knowledge due to mental illness is exculpating we need to look more closely at the mental capacities necessary to legal agency, and the way in which these capacities in particular can be compromised due to mental illness. H.L.A. Hart articulated these mental capacities in his discussion of "capacity responsibility" (Hart 1968). Hart viewed capacity responsibility as necessary for legal liability responsibility, which can result in a guilty verdict for a particular criminal act via attribution of *actus reus* and *mens rea*. The capacities Hart listed underpin a person's general ability to understand and conform one's behavior to rules, which he argued was a foundational requirement for the efficacy of law (Hart 1968). Hart noted that it is not fair or just to hold that a defendant has satisfied the mental state requirement for guilt unless that defendant has the capacity to recognize and behave in accordance with legal and moral rules. As stated above, the institution of law depends on persons being capable of understanding the rule of law and making choices which abide by legal rules; if no such persons existed, says Hart, then the institution of law would fail (Hart 1968). Similarly, if a person or class of persons cannot perceive the law as a reason to act and conform their behavior to it, the law fails as applied to that person or class because the law cannot influence their behavior—they fall outside of its reach (Hart 1968). Dana Nelkin and David Brink articulate a similar point in terms of persons having a "fair opportunity" to be law-abiding: only if a person has such a fair opportunity is it just to punish him or her when a crime is committed (Brink and Nelkin 2013).

Hart claims that capacities necessary to understanding the law's demands, and thus legal liability, include "understanding, reasoning, and control of conduct: the ability to understand what conduct legal and moral rules require, to deliberate and reach decisions concerning these requirements; and to conform to decisions when made" (Hart 1968, 227). Hart's list of capacities map fairly cleanly onto the M'Naghten test, which asks whether a defendant understands moral and legal rules such that they performed the criminal act with knowledge of its "nature and quality." It is important to note that Hart does not indicate that the incapacities that contribute to diminished mental capacity need to be related to a mental disorder; for example, Hart says juveniles are excused from criminal liability due to a lack of capacity responsibility. Most would agree that an eight-year-old can have the intent to kill another child, and act in a way that results in this child's death, yet not be deserving of criminal punishment.[5] As I will argue below, however, legal insanity, like the juvenile status excuse, captures a particular group of cases where persons suffer from severe incapacities for which they are unlikely to be culpable.

5.2 Mental Disorder Is "Weakly Relevant" to Legal Insanity

The structure of the M'Naghten test indicates that mental disorders are "weakly relevant" to legal insanity in that to be legally insane a defendant must have a mental disorder and certain symptoms related in some way to the crime they committed. In contrast, the presence of a mental disorder would be "strongly relevant" if a diagnosis itself indicated to the court that the defendant was legally insane. Even proponents of strong relevance admit that very few mental disorders are strongly relevant to legal insanity (Moore 2015). Psychosis, or schizophrenia, may be candidate disorders, with psychosis the most likely (Moore 2015).

---

[5] Law enforcement's intuitions regarding juvenile culpability sometimes fail them, as in this case where an eight-year old was arrested for murder: https://www.cnn.com/2015/11/11/us/alabama-boy-murder-charge/index.html.

Legal scholar Michael S. Moore embraces the position of strong relevance in part because he claims that a relationship of weak relevance doesn't work. If a mental disorder is strongly relevant to legal insanity, says Moore, then the fact that the defendant had that mental disorder at the time of the crime is enough to prove he is excused, because the mental illness is thought to compromise a person's moral and legal agency to such an extent that they are not responsible as a matter of course. On the other hand, if the incapacity caused by a mental disorder (such as a lack of moral knowledge) is thought to eliminate or diminish responsibility, we need not care how the incapacity came to be (e.g. due to a mental disorder or via some other means) (Moore 2015). In other words, if mental illness only excuses by causing a lack of moral or legal knowledge, it does no excusing work on its own. In this way a relationship of weak relevance indicates that the presence of a mental disorder is actually irrelevant to legal insanity (Moore 2015, 662).

Due to reasons discussed in Section 5.1, I don't think that a criminal court can infer from a mental disorder diagnosis alone that a defendant is legally insane. No mental disorder diagnosis, including psychosis, provides the court with enough information to assume a sufficient lack of moral knowledge at the time of the crime. In addition to a diagnosis, a defendant must provide evidence that he suffered from symptoms of that disorder in a way that undermined his capacity responsibility *at the time of the crime in way related to the crime*, such that the test for legal insanity is met (e.g. he lacked the ability to understand the nature and quality of his actions). Whether this defendant does indeed suffer from symptoms such that he lacked moral knowledge, the degree to which these symptoms were present at the time of the crime, and whether such symptoms were directly related to the criminal act for which the defendant has been arrested, must be determined by the fact-finding process. This process ought to include a psychiatric evaluation, and also evidence of the history of the defendant's disorder (e.g. what symptoms they were likely to be suffering from at the time the crime was committed, and the severity of those symptoms).

The illness most likely to underpin a successful plea of legal insanity in the US is a psychotic disorder, schizophrenia.[6] Persons with schizophrenia can suffer from a failure to understand the world as it actually is due to paranoia,

---

[6] Studies of various US states from the 1970s and 1980s reported that the insanity defense was pled in less than 1% of all felony cases, on average, across states, and seldom successful: defendants received a not guilty by reason of insanity verdict in only a fraction of that 1% of cases (Sales and Hafemeister 1984; Silver, Cirincione, and Steadman 1994). Some reports indicate over 80% of persons found legally insane are schizophrenic (Slovenko 1995).

hallucinations, and delusions, which are likely to generate false beliefs. They may also have cognitive deficits that make it hard for them to know that they are suffering from false beliefs about the world. They show decreased cognitive processing speed, difficulty ordering tasks, and difficulty multi-tasking, along with sometimes severe deficiencies in focusing attention and shifting attention between tasks, planning, and strategy capacity, and online use of working memory (Hirstein, Sifferd, and Fagan 2018). These capacities are very important to reasoning and are often termed "executive functions" by cognitive scientists. Importantly, although schizophrenic patients' psychotic symptoms often fluctuate, their executive deficits have been found not to track the severity of those symptoms; they persist even during periods of remission (Harvey et al. 2006; Krishnadas et al. 2014).

Although the science is still developing, evidence of executive functioning deficits in schizophrenia is important, because it means some persons with schizophrenia may not be able to identify and correct for false beliefs generated by hallucinations, delusions, and paranoia when these states occur. Verbal fluency tests have determined that schizophrenics have a heightened tendency to jump to conclusions (Ochoa et al. 2014); the higher the deficit in overall executive function in schizophrenics, the higher the tendency to jump to conclusions. Together, these two deficits—deficits that can result in proneness to false beliefs, and deficits related to an inability to identify beliefs as false and correct them—can in turn result in a lack of moral knowledge, because a person can hold false beliefs related to their criminal act, which can then result in an inability to understand the moral quality of that action.

This does not mean that persons with schizophrenia generally lack moral knowledge, however. Let's consider this diagnosis as a test case. Imagine a woman named Juanita who suffers from schizophrenia and lives in a community building for persons with serious mental disorders. Juanita has been very annoyed at her neighbor Ariel for years. Ariel is a loud neighbor, and her dogs bark incessantly. In addition, at several points Juanita has believed Ariel is spying on her and reporting back to the building's caretakers with the aim of getting Juanita evicted. On one occasion Juanita told her therapist she thought Ariel and her landlords wanted her dead. One evening Juanita slips a piece of paper she has lit on fire under Ariel's door. Ariel is asleep and dies from smoke inhalation.

Several questions must be answered to determine Juanita's eligibility for legal insanity. First, we would want to try to understand exactly what symptoms Juanita was experiencing when she killed Ariel, and to what extent. Was Juanita feeling very paranoid about Ariel spying on her? Did she believe Ariel was trying to have her murdered? That is, did Juanita have symptoms (such as paranoia and auditory hallucinations) generating false beliefs? And, if she did hold false beliefs related to the murder, did Juanita lack the executive capacity at the time to understand this belief may be suspect or even false?[7] We also might want to know whether Juanita is taking medication for her mental illness. Her diagnosis of schizophrenia may be completely irrelevant, or at least less relevant, to Juanita's responsibility if her symptoms, especially those that might general false beliefs, are well controlled by medication. If Juanita isn't taking medicine, we may use corroborating evidence from her health professionals, friends, and family, to attempt to attribute to Juanita some of the symptoms of schizophrenia that could result in a lack of moral knowledge. However, *why* Juanita isn't taking medicine may matter. If Juanita *should have* been taking medicine, but stopped taking it two weeks ago against her doctor's order, it could be that her symptoms are to some extent self-imposed, and thus not as exculpatory (even if they are severe and directly related to her criminal act).[8]

We can complicate the example further. Imagine that it becomes clear that Juanita *was* experiencing symptoms of schizophrenia such as hallucinations or delusions, but that they had nothing to do with Ariel: Imagine that after she went off her medication, Juanita grew sicker and became obsessed with the belief that Donald Trump was going to deport her; she even started wearing a special sweater that she believed would stop Trump from learning of her location. Juanita did not suffer from any delusions or hallucinations regarding her neighbor Ariel, only Trump; and she told several people in what seemed to be sincere moments of clarity: "If Ariel doesn't silence those stupid dogs, I'm going to do something to make her very sorry."

The point here is that it seems possible that even if Juanita is suffering from significant symptoms, such that she lacks moral knowledge in one

---

[7] Broome, Bortolotti, and Mameli (2010) discuss a similar example. They conclude that sorting out the interaction between the symptoms and a mental disorder and decision-making that leads to a crime is complex, and thus "general labels like 'mentally ill' are unlikely to be helpful in a context in which moral responsibility (or lack thereof) needs to be ascribed and punishment (or lack thereof) needs to be established" (Broome, Bortolotti, and Mameli 2010, 186).

[8] In this way we might "trace" Juanita's responsibility back to an earlier decision in the same way we blame drivers for the damage they cause whilst drunk, even if they are fully incapacitated at the time the damage is caused. They may be blameworthy based upon their earlier decisions to drink too much (Fischer and Tognazzini 2009). Instead of tracing, I have argued it makes more sense to consider capacity responsibility over time and hold persons responsible for earlier decisions made grounded in these capacities (Sifferd 2016).

context, she might commit a crime in a different context totally unrelated to these symptoms. On the M'Naghten test for legal insanity, if Juanita had sufficient mental capacity to understand the nature and quality of her criminal act of putting a lit piece of paper under Ariel's door (namely, that it was immoral and illegal),[9] she had fair opportunity to avoid the criminal act. Juanita should be excused only if she suffered from symptoms directly related to her moral knowledge regarding her murder of Ariel.

I have tried to show that even in the case of a person with severe schizophrenia who is having active symptoms at the time of their crime, their illness may be irrelevant to their responsibility. Since even a very serious mental illness like schizophrenia does not necessarily mean a person is incapacitated in a way that impacts legal culpability, we ought to be mindful that any mental disorder diagnosis, by itself, is insufficient to establish a lack of capacity responsibility (and thus legal insanity). Indeed, we ought not to be surprised if most mental illnesses are often irrelevant to responsibility. Mental illnesses such as OCD, depression, and phobias typically will not impact a person's capacity to understand the nature and quality of their actions. For example, typical symptoms of major depressive disorder include feeling sad, irritable, and excessively tired; and having difficulty concentrating or making decisions. None of these symptoms are likely to undermine a person's capacity to understand what conduct legal and moral rules require.

The argument above calls into question Moore's claim that mental illness is either strongly relevant to legal insanity or not relevant at all. Instead, mental illness is weakly relevant to legal insanity because some mental illnesses can cause symptoms that substantially affect moral knowledge, and thus undermine capacity responsibility in a localized way. At this point, one might again claim that it is the symptoms themselves—e.g. the presence of a hallucination and the inability to recognize it as such—that are relevant to responsibility. But causal historical information about the symptoms matters to responsibility. Importantly, persons are often not culpable for symptoms of a mental illness that can result in a lack of moral knowledge, where this may not be the case otherwise (e.g. if a person took LSD knowing that hallucinations were the likely result). Moore himself admits that there is an intuitive link between excuse and disability, grounded in the

---

[9] Although "ignorance of the law is not an excuse," knowledge of the immorality of an act serves as important information to citizens regarding whether it is also illegal (and not a justified or excused act) (Sifferd 2018).

idea that " ... [u]sually such disabilities are not our fault" (Moore 2015, 664). However, he says this intuition does not support an excuse, arguing instead that mental illness, if exculpatory, must provide an exemption. An excuse, Moore says, is for something someone does, a particular act, not for what they are (e.g. mentally ill). To serve as an excuse, there must be a particular relationship between the mental illness (and its symptoms) and the criminal act to be excused. For example, we might claim one is excused if the mental illness is causally related to the crime: We might say that "but for" the illness the act wouldn't have happened.

Below I will argue that the particular relationship that makes mental illness weakly relevant to criminal acts is not that "but for" the illness the act wouldn't have happened (a causal relationship often difficult to isolate and prove). Instead, as noted above, a mental health diagnosis provides important *epistemic* information to the court about the historical root of the local symptoms that may result in a person being legally insane with regard to a particular criminal act.

5.3 The Epistemic Utility of a Mental Disorder Diagnosis

"The crimes of legally insane offenders arise from a lack of understanding produced by a mental abnormality and thus they do not reflect culpable personal qualities and actions."
(Morse and Bonnie, 493)

Above I have tried to argue that on Hart's capacitarian account of legal responsibility we can treat the legal insanity excuse as operating in a "localized" way, where a mental disorder is weakly relevant to a defendant's responsibility because it can cause specific symptoms that may undermine a person's moral knowledge with regard to a particular criminal act. This approach rests upon a realistic understanding of the heterogeneous nature of mental disorders, as it is compatible with the idea that persons with the same mental disorder diagnosis can have different levels of capacity and incapacity at different times, in different environments, and with regard to different domains of action.

The presence of a mental disorder is important to legal responsibility, not because it tells the court that the defendant is legally insane but because it can signal to the court that a defendant may not be culpable for symptoms causing a lack of moral knowledge. The law contains many of these heuristic judgments—age as a rough proxy for cognitive maturity; marriage as a rough proxy for closeness of relationship—and they often don't work with 100% accuracy. However, they frequently serve as indispensable sources of information for the court.

In general, persons who are ignorant of the moral nature of their illegal action are not excused due to this ignorance. Our primary source of information about the content of the criminal law is societal moral rules (Husak 2016). Normally, ignorance of the law is not an excuse because we expect citizens to know these moral rules and the general content of criminal law. Placing upon citizens the responsibility to know the law is good policy because of its effects. Although, as noted above, in the US—and according to the MPC—the criminal law's primary purpose is state-imposed retribution for moral wrongs, the law also serves to accomplish forward-looking aims such as decreasing crime, and possibly, enhancing moral agency.[10] From this perspective, the principle that ignorance of the law does not excuse contributes to rule of law and social order by encouraging awareness of legal rules and law-abiding behavior.

It is fair for the law to require citizens to know the moral content of the law because adults have capacity responsibility, which grounds their ability to know and reflect upon moral and legal rules over time. As discussed above, a person must have capacity responsibility to qualify for legal liability and thus criminal guilt. Where a person has capacity responsibility but fails to understand the moral nature and quality of their act, the person is responsible because they have a legal obligation to use those capacities to reflect upon the moral and legal nature of their actions. On a capacitarian account persons are responsible *even if they do not in fact engage in this sort of reflection,* and thus are ignorant about the morality of an act that is criminal. Except in very rare circumstances,[11] the legal obligation to know the moral and legal demands of the law is enough to hold fully responsible a person who lacks moral knowledge in a way unrelated to a mental disorder.

As discussed above, persons with schizophrenia can suffer from delusions and hallucinations related to their actions, and can also lack the ability to

---

[10] See Manuel Vargas, *Building Better Beings: A Theory of Moral Responsibility* (Oxford University Press 2013) for an account of how forward-looking gains in moral agency might be grounded by backward looking assessments of responsibility.

[11] One example might be a very newly arrived immigrant from a country with very different moral rules in certain areas, like property rights or domestic relationships, who has not had a chance to become aware of her new countries' moral rules.

identify and correct for these beliefs given other things they know. If this is the case regarding beliefs and desires causally related to a criminal act, the person may be excused from that act via the M'Naghten test because they are not morally blameworthy for the act due to their lack of moral knowledge. Below I will compare this case with a hypothetical one, made famous by Susan Wolf (Wolf 1987), in which a man named "JoJo" lacks moral knowledge. Wolf argues that JoJo is morally insane (in an extended sense which is derivative of the legal sense but not identical with it), and thus not responsible for his acts, although it seems likely JoJo does not have a diagnosable mental disorder.

Here's the example in Wolf's own words:

> JoJo is the favorite son of Jo the First, an evil and sadistic dictator of a small, undeveloped country. Because of his father's special feelings for the boy, JoJo is given a special education and is allowed to accompany his father and observe his daily routine. In light of this treatment, it is not surprising that little JoJo takes his father as a role model and develops values very much like Dad's. As an adult, he does many of the same sorts of things his father did, including sending people to prison or to death or to torture chambers on the basis of whim. He is not coerced to do these things, he acts according to his own desires. Moreover, these are desires he wholly wants to have. When he steps back and asks, "Do I really want to be this sort of person?" his answer is resoundingly "Yes," for this way of life expresses a crazy sort of power that forms part of his deepest ideal.
>
> (Wolf 1987, 367–368)

Wolf intends this example to indicate that the "deep self" account of responsibility (Frankfurt 1971, Watson 1975) is not quite correct. This account claims that a person is responsible for actions governed by or answerable to second-order desires or values; but JoJo's action passes this test, and Wolf doesn't think he is responsible. JoJo is not responsible, she says, because JoJo doesn't really know the difference between right and wrong due to his bizarre upbringing. He does not really know which actions are morally wrong. JoJo has an "insane" deep self, and thus he is not responsible even though his first-order desires to torture and kill are endorsable by second-order values. (His values are that torture and killing are just fine, or even desirable.)

Whenever I teach the JoJo case I follow it up with a discussion of the true story of the "white power twins" raised in the US by their single mother who is a white supremacist.[12] The twins, named Lamb and Lynx, were home schooled and spent their summers touring KKK meetings as a band called Prussian Blue singing racist, hateful songs to entertain the meeting attendees. When the girls, at age 11, were first interviewed by the news program 20/20 in 2006, they were adamant that they believed the lyrics they were singing.[13] However, some years later one of the twins demanded to attend public high school. After spending time within the public school system, she recanted her previous racist views, saying she hadn't understood how terrible the views were given her limited environment growing up. Her sister also entered public school for her final year of high school, and sometime later, also seems to have rejected her mother's white supremacist views.

JoJo, and the latter real-life white power twins—before they recanted— seem to lack moral knowledge due to extreme environments where they were not exposed to typical societal values. Imagine that the second twin, Lynx, never went to public school, continued to live with her mother, and had little interaction with persons outside the KKK. Next imagine that as adults JoJo and Lynx commit crimes related to their upside-down value systems. If, due to a lack of moral knowledge, JoJo and the twin are properly considered legally insane, then my argument that the presence of a mental illness is important to legal insanity may be wrong. But I do not believe this is what these cases tell us. Let's look at the white power twin case a bit closer.

When the girls are first interviewed on ABC's 20/20 program, they are still quite young. At this age, the girls are not fully responsible for their actions because they are juveniles. They lack moral knowledge, and this lack is due to factors outside their control, including their immature cognitive systems. They are still very impressionable, and because they are homeschooled, their indoctrination into the belief system of the KKK had not been challenged by a contrary value system. This contrary value system would have been present within public school and other parts of the girls' community if they had been living more normal lives. When she entered public school, Lamb was exposed to the values more typical of our liberal democracy, and she recanted her racism fairly soon after. At this point, the twin was more mature, and thus more capable of reflecting upon her family's value system. Later, it seems Lynx had the same sort of experience. Among the

---

[12] The Southern Poverty Law Center has an entry on Lamb and Lynx's mother April Gaede that describes her ideology (https://www.splcenter.org/fighting-hate/extremist-files/individual/april-gaede).

[13] For a summary of the twins' earlier interview and an update on their change of heart, see the ABC newscast recorded 7/20/2011: https://www.youtube.com/watch?v=ULsTm5VR73c.

underappreciated functions of public schooling (and schooling in general) are the functions of exposing children to values other than those of their immediate family; exposing them to the values commonly held within their community; and helping calibrate the child's moral compass so that it does not run radically contrary to those of the child's larger society despite their parent's idiosyncrasies.[14]

Both twins eventually became exposed to their larger community and rejected their mother's values. But again, let's imagine that Lynx did not go to public school or recant racism, and at the age of 22 this twin was arrested for burning a cross on the front yard of a local African American family. What would we think about her claiming she didn't understand her action of burning a cross on the neighbor's lawn is wrong? Imagine Lynx claims burning the cross is a morally correct act, and that she had the strong sense that God told her to perform the action, and that she experienced "signs" that he was encouraging her.

It seems to me that even in this case Lynx is legally responsible for this act despite (perhaps) her lacking moral knowledge. At 22, we can now presume she has capacity responsibility: she is old enough to have developed the capacity to reflect upon her beliefs and actions considering the moral values of her society and the law. Further, she is now not so isolated by her mother to lack exposure to those values; she knows she has the ability to leave home and socialize and work outside of the white supremacist movement, because she has seen her sister do this. Lynx is likely to have experienced other value systems, not just through her sister, but also via the media, magazines and books, and TV. She is likely to know burning crosses is considered both morally wrong by most and is illegal. The law requires her to know its moral demands, even if this requires some effort on her part. Due to her capacity responsibility and her ability as an adult to seek out societal values, the twin now has a fair opportunity to avoid breaking the law (Brink and Nelkin 2013). When she burns a cross on someone's lawn, she isn't legally insane or otherwise excused even in the unlikely event that she remains ignorant that her act is immoral or illegal. She's racist, and guilty of a crime.

One might worry that this twin had less of a fair opportunity to be law-abiding than some other persons raised differently. This is true, in the same way that persons raised very poor in a violent neighborhood may find it harder to be law-abiding. But the law cannot parse cases along fine-grained

---

[14] Anneli Jefferson commented that it may also be the case that public schooling can weaken strict moral standards a student has been exposed to at home.

lines of difficulty. Except in the most bizarre of circumstances (see my final discussion of JoJo below), people are very likely to know killing and stealing are at least perceived to be wrong and illegal by the majority in society and the law. Note that citizens don't have to internalize societal values fully to be responsible; as Hart noted, not all citizens with full capacity responsibility will internalize the law and its moral counterparts as a reason to act (Hart 1968). To have the capacity to understand the nature and quality of one's act one needn't agree with society's moral perception of it, one only needs the capacity to understand that perception. Citizens with capacity responsibility who fail to internalize the moral tenants underpinning law will generally still have had the opportunity to review the values they hold in light of societal values, and the capacity to do so. Thus, when they act contrary to those values they will know society is likely to hold them accountable.

As a practical matter, the law utilizes fairly coarse categories: there is a presumption of capacity responsibility at age 18 and the law assumes this capacity operates against a background of somewhat "normal" acculturation so that citizens can take note of the law's demands (via detection of societal values) and conform to them. Although I would argue full capacity responsibility develops a bit later (Fagan, Sifferd, and Hirstein 2020) the assumption of acculturation is reasonable. As the twins cases shows, even children raised within a family that embraces abhorrent moral values is likely to encounter mainstream values in school or in their community. The law asks us to adhere to the moral norms of our society, reflected in the criminal law, and once we are adults we can ensure our moral values are not bizarre by seeking out information about the moral views of others: in school settings or through books, TV, and other media, or more directly by talking to people who are different from ourselves or participating in activities with them. In this way we can make sure we at least can detect, and possibly understand, the moral values of our society. On this view, willfully siloing oneself away from the moral values of the majority is a culpable act and can lead to criminal responsibility even in cases where we lack moral knowledge. And, by imposing these expectations on citizens, the law may have the forward-looking effect of supporting and cultivating persons' sensitivity to moral considerations (Vargas 2013).

By adulthood, persons with capacity responsibility can begin the process of diachronic review of the moral values they were taught as a child, and self-authorship with regard to these moral values. One who learns their values run contrary to society's can take steps to become law-abiding even if they don't fully adopt these values. Adina Roskies has argued this sort of diachronic self-authorship grounds responsibility in a way that preserves agency (Roskies 2012). We can deliberately shape our future selves, says Roskies, by manipulating our mental content, including our values, in ways that have foreseeable consequences, and because we have such diachronic control we are in a "very real sense responsible for who we are" and our behavior (Roskies 2012). Even if the values we were raised with are deeply felt—e.g., a person raised with racist values feels fear when they are confronted with persons of another race—we can reflect upon these values and manipulate ourselves to stop from acting in a racist way. Examples of diachronic self-interventions include making commitments to future moral and law-abiding behavior or setting overarching policies for behavior ("I will not cross the street when I see someone of another race; I will look them in the eye and smile"), and engineering one's environment so that moral and law-abiding behavior is easier to perform.[15]

Let's return to JoJo. JoJo's environment is so bizarrely constrained that it seems possible he is never really exposed to values other than those held by his tyrant father (e.g. the typically held moral values represented in a criminal law system like ours). He is removed from the possibility of the normal process of acculturation, and bizarre values are impressed upon him in highly violent and stressful circumstances. Given all this, I think the JoJo case is more one of brainwashing than insanity, precisely because he is completely isolated from more typical moral norms due to no fault of his own. Some have argued that extreme brainwashing may include superimposed or implanted *mens rea*, such that the criminal intent that grounds an action is not the actor's own (Delgado 1978). This may result in negated *mens rea* and acquittal or in a claim of diminished mental capacity, where the defendant is only partially responsible.

Although I do not think the JoJo case is one of legal insanity, it is helpful to my argument that mental disorder is weakly relevant to legal insanity because it exposes the background against which the plea of legal insanity operates. Generally, persons are assumed to have been exposed to the community values echoed in the criminal law. Persons who fail to adopt those values, then, are held responsible when they violate them. A person who does not understand right from wrong is typically in this state in a culpable way—they have had the opportunity to understand an act is wrong according to societal

---

[15] These tactics may not wipe away racist feelings completely, but can stop racist action. There is some evidence that it is awareness of one's racism that allows one to stop acting racist, not the removal of implicit racist views (Monteith 1993).

values and the law via the normal processes of inculturation. Capacity responsibility rests upon a diachronic process of development of moral agency within a societal context.

The presence of a serious mental illness represents one of the rare cases where a person might non-culpably lack moral knowledge regarding the nature and quality of an act. This is because a very serious mental illness—of the sort that can ground a successful insanity plea—can provide a person with false information about the circumstances within which they act via hallucinations and delusions, and also deny the person the ability to identify and correct these false beliefs. If someone believes their children are doomed to hell and the only way they can save the children is to drown them; or that their neighbor is a government spy with the aim of killing them, these beliefs taint that person's moral understanding of their circumstances, and actions they commit based on this understanding. Although the science on this is not yet clear, schizophrenia may also result in executive deficits that deny persons the ability to understand the moral weight of her actions given societal norms. That is, a person with schizophrenia who believes their neighbor plans to kill them may jump to the conclusion that they need to act first, without the rational capacity to evaluate this plan of action against other plans that may be more reasonable given societal norms. Thus, the presence of a mental illness can provide crucial information to the court about the nonculpable nature of a defendant's incapacity to understand the nature and quality of a particular act.

5.4 Conclusions

In this chapter I have argued that using a capacitarian account of legal responsibility we can treat the legal insanity excuse as operating in a "localized" way, where a mental disorder is weakly relevant to a defendant's responsibility because it can cause specific symptoms that may undermine a person's moral knowledge with regard to a particular criminal act. This approach is compatible with the idea that persons with the same mental disorder diagnosis can have different levels of capacity and incapacity at different times, in different environments, and with regard to different domains of action. Some persons with a serious mental disorder such as schizophrenia may generate false beliefs about the world, and have the inability to correct for these false beliefs. Where these symptoms are related to a criminal act, a person may be said to lack moral knowledge regarding this act in a non-culpable way. Generally, persons who lack moral knowledge, but have capacity responsibility, are fully responsible for their actions. Legal insanity is a legal category created to excuse persons who suffer from a lack of moral knowledge due to a severe illness over which persons often have very little control. If there are other types of cases where a person is not culpable for their lack of moral knowledge—such as may be the case with poor, brainwashed JoJo—they may also be (partially) excused, but via other legal means.

References

American Law Institute (ed.). 2007. Model Penal Code: Sentencing, Tentative Draft No.1, Part I. General Provisions. The American Law Institute, April 9, 2007.

Brink, D., and D. Nelkin. 2013. "Fairness and the Architecture of Responsibility." *San Diego Legal Studies Paper:* 13–132.

Broome, M., L. Bortolotti, and M. Mameli. 2010. "Moral Responsibility and Mental Illness: A case study." *Cambridge Quarterly Healthcare Ethics* 19 (2): 179–187.

Delgado, Richard. 1978. "Ascription of Criminal States of Mind: Toward a Defense Theory for the Coercively Persuaded ('Brainwashed') Defendant." *Minnesota Law Review* 63.

Fagan, T., Sifferd, K. & Hirstein, W. 2020. *Juvenile Self-Control and Legal Responsibility: Building a Scalar Standard.* In Alfred Mele, (ed.), *Surrounding Self-Control.* Oxford: Oxford University Press.

Fischer, J.M., and N. Tognazzini. 2009. "The Truth about Tracing." *Nous* 43 (3): 531–556.

Frankfurt, H. 1971. *Freedom of the Will and the Concept of a Person,* Journal of Philosophy 68:1, 5–20.

Hart, H.L.A. 1968. *Punishment and Responsibility: Essays in the Philosophy of Law.* Edited by Oxford University Press. Oxford: Clarendon Press.

Harvey, P., Koren, D., Reichenberg, A., and Bowie, C. 2006. *Negative Symptoms and Cognitive Deficits: What Is the Nature of Their Relationship?* Schizophrenia Bulletin 32:2, 250–258.

Hirstein, W., K. Sifferd, and T. Fagan. 2018. *Responsible Brains: Neuroscience, Law, and Human Culpability.* Cambridge, MA: MIT Press.

Husak, Douglas. 2016. *Ignorance of Law: A Philosophical Inquiry.* New York: Oxford University Press.

Jefferson, Anneli, and Katrina Sifferd. 2018. "Are Psychopaths Legally Insane?" *European Journal of Analytic Philosophy* 14 (1): 79–96.

King, M., and J. May. 2018. "Moral Responsibility and Mental Illness: A Call for Nuance." *Neuroethics* 11 (1): 11–22.

Krishnadas, R., Ramanathan, S., Wong, E., Nayak, A., and Moore, B. 2014. "Residual Negative Symptoms Differentiate Cognitive Performance in Clinically Stable Patients with Schizophrenia and Bipolar Disorder." *Schizophrenia Research and Treatment*, vol. 2014. https://doi.org/10.1155/2014/785310.

Monteith, M.J. 1993. "Self-Regulation of Prejudiced Responses: Implications for Progress in Prejudice-Reduction Efforts." *Journal of Personality and Social Psychology* 65 (3), 469–485.

Moore, Michael. 2015. "The Quest for a Responsible Responsibility Test: Norwegian Insanity Law after Brievik." *Criminal Law & Philosophy* 9 (4): 645–693.

Morse, S. 2015. "Neuroscience, Free Will, and Responsibility." In *Free Will and the Brain: Neuroscientific, Philosophical, and Legal Perspectives*, edited by W. Glannon. Cambridge: Cambridge University Press.

Morse, S. 1994. *Culpability and Control*. University of Pennsylvania Law Review 142, 1587–1660.

Morse, Stephen, and Richard Bonnie. 2013. "Abolition of the Insanity Defense Violates Due Process." *The Journal of the American Academy of Psychiatry and the Law* 41: 488–495.

Ochoa, S., Haro, J.M., Huerta-Ramos, E., Cuevas-Esteban, J., Stephan-Otto, C., Usall, J., Nieto, L., and Brebion, G. 2014. "Relation between jumping to conclusions and cognitive functioning in people with schizophrenia in contrast with healthy participants." *Schizophrenia Research* 159:1, 211–217.

Roskies, Adina. 2012. "Don't Panic: Self-authorship without Obscure Metaphysics." *Philosophical Perspectives* 26.

Sales, B. & Hafemeister, T. 1984. "Empiricism and Legal Policy on the Insanity Defense." In Teplin, L., (ed.), *Mental Health and Criminal Justice*. Sage Publications, 253–278.

Sifferd, K. 2016. "Unconscious Mens Rea: Lapses, Negligence, and Criminal Responsibility." In *Philosophical Foundations of Law and Neuroscience*, edited by D. Patterson and M. Pardo. Oxford: Oxford University Press.

Sifferd, K. 2018. "Ought Ignorance of Criminal Law (and Its Morality) Excuse? An Essay Review of Husak, D. *Ignorance of Law: A Philosophical Inquiry.*" *Jurisprudence* 9 (1): 186–191.

Silver, E., Cirincione, C. and Steadman, H.J. 1994. "Demythologizing inaccurate perceptions of the insanity defense." *Law and Human Behavior* 18:1, 63–70.

Slobogin, C. 2000. "An End to Insanity: Recasting the Role of Mental Disability in Criminal Cases." *Virginia Law Review* 86 (6): 1199–1247.

Slovenko, R. 1995. *Psychiatry and Criminal Culpability*. Oxford: Wiley.

Vargas, Manuel. 2013. *Building Better Beings: A Theory of Moral Responsibility*. Oxford: Oxford University Press.

Watson, G. 1975. "Free Agency." *Journal of Philosophy* 72:8, 205–20.

Wolf, Susan. 1987. "Sanity and the Metaphysics of Responsibility." In *Responsibility, Character, and the Emotions*, 46–62. Cambridge: Cambridge University Press.

6

Scrupulosity and Moral Responsibility

Jesse S. Summers and Walter Sinnott-Armstrong

Adam (a pseudonym) is a patient who has obsessive thoughts that a "grocery store cashier may have made an error in [his] favor." The patient then engages in a compulsion: "Receipts are taken home and laboriously checked item by item, even if totals involve hundreds of dollars. Some receipts are kept for months, with the patient hoping that the urge to check will eventually decline and he can throw them away."

(Ciarrocchi 1995: 42)

Imagine that you are Adam's child. You try to talk to your father about an important personal issue, but his attention is clearly focused instead on receipts that he had already checked dozens of times. You feel frustrated and angry, but are those appropriate reactions? Is he acting immorally by not supporting you? Or would you feel only pity for him? You know that he loves you, wants to help, and would listen intently if he were not so obsessed. His obsession hurts him as well as you.

Compare another case where the issue is sex instead of honesty:

Linda (another pseudonym) reports, "I am troubled with bad thoughts and desires. I am afraid to bathe or brush against my breast for fear I will feel sexual pleasure. I have harmful and envious thoughts about others. I am afraid to watch TV because of the bedroom scenes. I'm even afraid I'm abusing my health by getting so upset about these things and maybe that is a sin also. My common sense tells me that these are either no sin at all or, at most, venial sins, but I'm never sure, so I stay away from Holy Communion. When I see so many people receiving Communion, I want so badly to go, but I can't because I feel so unworthy." (Santa 2007: 137)

Jesse S. Summers and Walter Sinnott-Armstrong, *Scrupulosity and Moral Responsibility* In: *Agency in Mental Disorder: Philosophical Dimensions*. Edited by: Matt King & Joshua May, Oxford University Press.
© Jesse S. Summers and Walter Sinnott-Armstrong 2022. DOI: 10.1093/oso/9780198868811.003.0007

Imagine that you are Linda's spouse, and she does not bathe or brush for a week. She is constantly distraught and never agrees to do anything with others. Her obsession is undermining your marriage and your lives. You cannot help but feel frustrated, but she is hurt as much as you—maybe more. She loves you and wishes she could overcome her obsession, but she can't. If you were Linda's spouse or Adam's child, would you feel anger or pity or both? Which feeling would be justified, and why?

These real cases exemplify a condition called Scrupulosity. Scrupulosity raises profound questions about the nature of mental illness, moral judgments, and moral responsibility. We will begin by explaining this condition and arguing that it is a mental illness. Then we will discuss how it distorts moral judgments and thereby undermines or at least reduces moral responsibility. We will also show how this condition challenges popular deep-self theories of moral responsibility.

6.1 What Is Scrupulosity?

Scrupulosity is a form of obsessive-compulsive disorder (OCD) that focuses on moral or religious obsessions or compulsions. Central to all forms of OCD is the underlying anxiety that forms and sustains the person's obsessions and that their compulsions are intended to reduce. Therefore, we will focus much of our discussion on this underlying anxiety that is central to all forms of OCD, including Scrupulosity.

Of course, the most visible symptoms of OCD are not its underlying anxiety, but its obsessions and compulsions. Obsessions are persistent intrusive unjustified thoughts that invoke and respond to underlying anxiety. These thoughts could be beliefs, desires, images, or perhaps other mental states. They are persistent when they last for an extended period of time or return with regularity. They are intrusive because they conflict with what the person takes to be her underlying concerns and goals, so she does not want to have the thoughts at all. By contrast, persistent but justified thoughts, such as fears of real and recurring dangers, would not count as obsessions, nor would persistent, intrusive, and unjustified thoughts that are not associated with any anxiety.

Whereas obsessions are mental events or states, compulsions typically involve actions that the person with OCD performs in response. These actions might include bodily movements, such as washing hands or locking doors, but they also might be purely mental, such as when an individual with

OCD must count the number of rods in a fence while walking by for fear that something bad will otherwise happen. These compulsive actions are in response to the person's obsessions and serve—at least temporarily—to reduce their anxiety.

Scrupulosity is generally like other forms of OCD, except that its obsessions and compulsions are moral or religious. In our opening examples, Adam obsesses about the possibility of cheating a store or at least failing to pay what he owes, and he compulsively counts and recounts his receipts to reassure himself that he has not failed to remedy a mistake in his favor. Linda's "bad thoughts and desires" are her obsessions, and her compulsions include refusing to take showers or communion, intentionally avoiding an action that she wishes she could do.

Although Scrupulosity shares these defining features with other forms of OCD, it also has three other characteristic features.

First, people with Scrupulosity typically exhibit *moral perfectionism*. This means that they have extremely high moral or religious standards, at least for themselves. Most of us believe that we should do something to help those less fortunate than we are, but a person with Scrupulosity might work constantly on behalf of those in need out of a sense that he is otherwise morally failing them. Similarly, most of us believe that we should pay stores all of the money we owe, but many of us don't check our receipts at all, and almost none of us feels a need to check our receipts more than once to be sure we haven't failed to remedy a mistake in our favor. The moral standards patients with Scrupulosity apply to themselves are familiar to all of us, but patients strengthen these common moral standards at least for themselves and hold themselves to be moral failures if they cannot reach such exacting standards.

Second, many people with Scrupulosity also exhibit *moral thought–action fusion*. In other words, they treat having thoughts about immoral behaviors as morally equivalent to actually performing those immoral behaviors. A person with Scrupulosity imagined having sex with Jesus every time she saw him lightly clothed on a crucifix, and she thought that merely having the idea of such an act was just as bad or nearly as bad as performing the act in reality—even though she was not worried that she was going to act on her thoughts, because she knew that she couldn't. It's not uncommon to worry about whether our thoughts are good or whether they reveal something bad about ourselves, but moral thought–action fusion goes beyond these common moral judgments by seeing immoral acts as no worse (or not much worse) than thinking about immoral acts. To this extent, they fuse—or fail to distinguish morally between—having a thought and acting on it.

A third feature that often characterizes Scrupulosity is *chronic doubt and intolerance of uncertainty*. People with Scrupulosity find it hard to be reassured about their doubts, both about moral issues and in general, and they find it anxiety provoking to be unable to settle moral uncertainties. They go through their lives constantly doubting whether they are good enough and whether they have done enough to meet their perfectionist standards of morality.

Scrupulosity is defined by its obsessions and compulsions and is characterized by perfectionist moral standards, moral thought–action fusion, chronic doubt, and intolerance of uncertainty. Of course, many people can be "a little OCD" about moral issues and have some scrupulous traits that fall short of a clinical diagnosis. They might, for example, worry about their moral obligations or check whether they've done the right thing. They might have high moral standards for themselves, and even insist that they should maintain pure thoughts. The difference between someone who has Scrupulosity and someone who has all these traits but would not be diagnosed with Scrupulosity can be a matter of degree, and there is no doubt much of interest to say about those who would not be diagnosed. But we will focus here only on clinically significant cases of Scrupulosity.

6.2 Is Scrupulosity a Mental Illness?

Even though OCD is a recognized mental disorder, and Scrupulosity is a form of OCD, it still might seem to remain an open question whether or not Scrupulosity is a mental disorder. People with contamination OCD wash their hands more than they need for the sake of cleanliness. Others with OCD check locks on their homes more than they need for security. But what does it mean to say a person is "excessively" worried about being a bad person or committing an immoral act? Isn't being worried about doing the right thing a sign of a good person?

Nonetheless, Scrupulosity does fit not only the criteria for OCD but also the general definition of mental disorder in *Diagnostic and Statistical Manual of Mental Disorders* (DSM-5):

A mental disorder is a syndrome characterized by clinically significant disturbance in an individual's cognition, emotion regulation, or behavior that reflects a dysfunction in the psychological, biological, or developmental processes underlying mental functioning. Mental disorders are usually associated with significant distress or disability in social, occupational, or other important activities. An expectable or culturally approved response to a common stressor or loss, such as the death of a loved one, is not a mental disorder. Socially deviant behavior (e.g., political, religious, or sexual) and conflicts that are primarily between the individual and society are not mental disorders unless the deviance or conflict results from a dysfunction in the individual, as described above. (DSM-5, p. 20)

We will not compare the many alternative definitions or defend this particular definition here (see Singh and Sinnott-Armstrong 2015), but we will ask whether Scrupulosity meets the conditions in this definition.

Scrupulosity is clearly a syndrome insofar as it combines a variety of symptoms discussed in the previous section. These symptoms include disturbances in cognition (worry about one's moral behavior or thoughts), in emotion regulation (especially anxiety), and behavior (in compulsions). These features reflect dysfunctions in psychology—and perhaps in biology and development—because one cannot function normally with such high and persistent levels of anxiety and guilt. People with Scrupulosity such as Adam and Linda often find themselves unable to do what they most want to do, trapped by their own anxiety. Although Scrupulosity could begin after a loss or stressor, its symptoms go far beyond what is culturally expected. The behavior of people with Scrupulosity is clinically significant not solely because other people disapprove of it but because of the anxiety, guilt, distress, and disability in patients as well as harmful effects on others. Thus, all of the conditions in the definition of mental disorder in DSM-5 are met by Scrupulosity.

The same cannot be said of moral saints, which is why we can distinguish the mental illness of Scrupulosity from the rare case of being morally exemplary. Consider Zell Kravinsky, who gave most of his fortune to charity and donated a kidney to a stranger, among other saintly acts. His behaviors were extreme and unusual, but they were not motivated by anxiety, they did not lead to distress or disability, and there was no sign of dysfunction in his psychological, biological, or developmental systems. Moreover, his behaviors were culturally approved and even praised by many. His wife was reportedly angry about his kidney donation, perhaps because of its impact on her, but the DSM definition excludes "conflicts that are primarily between the individual and society." Thus, the definition of mental illness that captures Scrupulosity would not also entail that Kravinsky has a mental disorder. The same points apply to most other moral saints, unusually extreme altruists, and morally exemplary people.

6.3 Moral Judgments

We have already seen that people with Scrupulosity make unusual moral judgments. Their moral judgments are perfectionist, equate thought and action, and are sometimes fueled by chronic doubts and intolerance of uncertainty about what they should do and whether they have done enough. Despite these differences in degree from what most people believe about their moral obligations, their judgments still have much of the same content as ordinary moral judgments: people with Scrupulosity try to be honest, harm others very little if any, and help the needy. They do not, for example, judge that they have moral obligations to count blades of grass or stand on their heads. In content, the judgments that people with Scrupulosity make about morality are not wildly different from moral judgments of those without.

That they share in the same general content, however, isn't enough to show that they're making genuinely moral judgments. For one thing, their judgments are driven by anxiety in a way that can distort them. This distortion isn't enough to change their content from the moral to the non-moral, but it is enough to raise questions about whether the content reflects genuinely moral concerns.

Consider two ways in which the anxiety that underlies Scrupulosity can make a difference to the person's judgments. First, people with Scrupulosity might sometimes make quite ordinary moral judgments (like judging that they need to help the poor) that prompt excessive or persistent anxiety, which then lead to further moral judgments, such as that they are required to help even more needy people and maybe to apologize for not doing more to help the poor. Alternatively, people with Scrupulosity might sometimes feel strongly or persistently anxious, and, as a way of rationalizing this ever-present anxiety, they conclude that they are regularly committing moral wrongs. The anxiety-induced moral evaluation of themselves then informs the judgments they make about what they should do, e.g., that they should apologize yet again for a wrongdoing that they've apologized for three times already. Actual cases likely involve anxiety running in both directions: from judgment to anxiety and from anxiety to judgment.

In many cases, people with Scrupulosity might feel anxious and interpret that as evidence that they've done something wrong, then notice something that they indeed might have done wrong. The anxiety has directed their attention to make a judgment they would not otherwise have made—and perhaps needn't make. But the judgment might be a perfectly legitimate one, however it came about, and they now are genuinely worried about what they think they've done. They resolve to do something to make things better, which will also make the anxiety go away. But their anxiety doesn't go away (or doesn't stay away) when they try to make things better, so they become more anxious or try more extreme ways of remedying their perceived wrongdoing. They shift their moral standards in the direction of perfectionism in order to make sense of why they feel anxious to this high degree despite efforts to resolve it, and they make further, subsequent judgments on the basis of these higher standards. Thus, their underlying anxiety and their anxiety-inducing judgments reinforce each other.

What's notable here is that these judgments about what to do are shaped to a great degree by a desire to reduce their anxiety, not simply a desire to do the right thing. People without anxiety disorders might reasonably be anxious about having done the wrong thing and might even take their anxiety to be a sign that they have done something wrong. Mild anxiety can even have some benefits, e.g., focusing attention on what the person takes to be most important in a complicated situation (Kurth 2018). In contrast, the kind of anxiety in Scrupulosity and anxiety disorders is extreme in both intensity and persistence. Such extreme anxiety can have distorting effects on the moral judgments. It can, for instance, lead a person to focus on those features of a situation that are most relevant to reducing or rationalizing the extreme anxiety, and those might not be the most important moral features. For example, if a patient is worried that not washing her hands has made a friend sick, she is focusing on something that she can do something about, but she is not focusing on the many other ways in which her friend might become sick, some of which are far more important than whether she washed her own hands.

Also, extreme anxiety can lead one to maintain that focus and one's related judgments that one has done something wrong, which makes the anxiety-driven judgment inflexible and unresponsive to ways of remedying the situation. If Pat apologized to Sam for saying something that Pat afterwards feared was offensive, and Sam accepted the apology or even told Pat that it was not in fact offensive, then Pat might still be anxious. Pat might take this persistent anxiety as evidence that something about the apology was flawed: Sam didn't understand the slight, so Sam forgave too easily, or Sam didn't really forgive despite saying so. As long as the anxiety persists, Pat can continue to take it as evidence that something morally problematic remains, which makes Pat's subsequent judgments inflexible.

So, while the presence of anxiety itself is not enough to taint a moral judgment, since ordinary anxiety can occur even in genuine moral judgments, extreme (intense or persistent) anxiety can have distorting effects on a moral judgment. If the judgment is distorted enough—as it easily can be in the cases of anxiety disorders—then it might be more appropriate to explain the putatively moral judgment as a response to the extreme anxiety than as a response to any truly moral concerns. The judgment is based on the judge's own welfare rather than the welfare or rights of others. Therefore, it might no longer be a genuinely moral judgment at all, at least if genuine moral judgments are moral insofar as (and because) they respond primarily to moral concerns like the welfare and rights of others.

Finally, anxiety can lead a person to make moral judgments that respond to some moral reasons considered narrowly, but not to the totality of moral reasons considered more broadly. Imagine that a parent is anxious about violating the bodily autonomy of an infant by having it vaccinated. Violating bodily autonomy is a moral concern, even for infants. However, if the parent makes a judgment based solely on that consideration and ignores the large and potentially deadly risks for the child and perhaps also for hundreds or thousands of other children now and for years to come, then it becomes clear that the parent's judgment should not have been made so narrowly. A moral judgment should instead consider all the obviously relevant moral features or, at a minimum, should not be limited just to those features that make one feel particularly anxious. In this way, too, anxiety can lead one to make moral judgments that are, at best, highly distorted and, if the distortion is significant enough, not moral at all.

6.4 What Is Moral Responsibility?

Since the moral judgments of those with Scrupulosity are, at a minimum, distorted, one must wonder whether people with Scrupulosity know whether what they are doing is morally right or morally wrong. At the extreme, if they cannot know moral right from wrong, they are less morally responsible than they would otherwise be and perhaps not morally responsible at all. To see whether this is true of those with Scrupulosity, we first need to understand the nature of responsibility.

An agent can be responsible for a certain action or for a certain consequence of the action. People with Scrupulosity almost never cause severe harms or commit serious crimes, such as murder, burglary, or fraud.

Nonetheless, people with Scrupulosity do cause less severe harms. We imagined some harms in the cases of Adam and Linda, and people with Scrupulosity have lied to their parents, surreptitiously stalked a former girlfriend, and annoyed their coworkers. Even though these are relatively minor harms, we can still ask whether they are responsible for even minor harmful actions or consequences of their Scrupulosity.

To answer this question, we need to clarify which kind of responsibility is at stake. Gary Watson and David Shoemaker helpfully distinguish two kinds of responsibility: attributability and accountability (Watson 2004; Shoemaker 2011, 2015). A character defect is attributable to an agent on the basis of an action when and only when that action shows that the agent has that character defect. However, we do not always view such condemnable agents as accountable for their own defects. Watson (2004: 235–52) discusses the case of Robert Harris, who committed unspeakable crimes, but who also was abused so severely as a child that most people do not think of him as fully responsible for turning into the monster that he was. In short, he was attributable but not accountable for his deeds and the resulting harms.

This distinction is useful for understanding Scrupulosity, because those with Scrupulosity seem to have attributability, though their accountability is a separate matter. There's no question that Adam is honest. That character trait is attributable to him on the basis of his actions. What is not so clear is whether Adam is accountable either for his own character trait of honesty or for the resulting actions of reviewing his receipts repeatedly or for the consequences of those actions, such as hurting the feelings of his children. To simplify our discussion, let's ask only whether those with Scrupulosity are responsible in the accountability sense.

Intuitively, Adam seems less responsible than would be an agent with no mental illness who performs similar actions with similar consequences. If an Adam without Scrupulosity had spent his evenings reorganizing his coin collection or watching football, rather than helping his children study for an important test in school or attending their performance in a musical concert, then he would be more accountable for the harms (including disappointment, embarrassment, and failure on the test) that he caused to his children than he would be for doing the same things as a result of his Scrupulosity. They (and perhaps observers) would seem more justified in being angry at unscrupulous Adam, and such justified anger can be a sign that he is accountable (Shoemaker 2015). In contrast, Adam with Scrupulosity is more like a parent who cannot help his children study or attend their concert because his back was injured through no fault of his own, so it would be very painful for him to help or attend. Both a mental disorder and a physical ailment would diminish both how appropriate it is to feel angry with Adam and how responsible Adam seems to be for the very same acts. Thus, in what we expect is an ordinary and intuitive view, an agent who causes harms by performing actions as a result of Scrupulosity is less responsible than someone who does the same acts and causes the same harms without Scrupulosity.

We do not deny that Adam with Scrupulosity remains responsible for these harms to some extent. He wasn't kidnapped or in a coma, conditions that would remove his responsibility entirely. But then the challenge is that, because his responsibility is diminished but not removed by his Scrupulosity, we need to explain both why such agents are not fully responsible and also why they remain responsible to some degree.

6.5 Incompatibilist Theories of Moral Responsibility

To explain this intermediate position, we obviously need to rely on an account of responsibility that allows responsibility to vary in degrees.

Some theories do not allow responsibility to vary in degrees, because they deny that any agent is ever responsible to any degree. For example, hard determinism and hard incompatibilism both state that the physical world is deterministic and that this precludes moral responsibility (Pereboom 2001). On such views, Adam with Scrupulosity is not responsible for not helping his children, just as he wouldn't be if he were kidnapped. Indeed, according to these views, Adam would be no more responsible if he simply refused to help his children for no reason, since all of his choices are determined by the past and the laws of nature. We find this indiscriminate denial of responsibility implausible and unhelpful, so we'll set these theories aside.

In contrast, libertarian theories of responsibility hold that determinism is incompatible with moral responsibility, but some human acts are not determined (van Inwagen, O'Connor, Clarke, Ginet, and Kane in Kane 2002). These views usually allow that some agents are responsible for some acts and not others, but they never provide any practical way to tell whether a particular agent is or is not determined or responsible for a particular act in a realistic situation. Moreover, they imply that any agent who is responsible at all is fully responsible. After all, a particular action is either determined or not, as well as caused by the agent or not. None of these notions admit of degrees.

These incompatibilist theories of responsibility, thus, cannot explain why people with Scrupulosity are partly but not fully responsible. For this reason, we will focus henceforth on compatibilist (and semi-compatibilist) theories of responsibility, on which determinism is compatible with moral responsibility.

6.6 Deep-Self Theories of Moral Responsibility

One of the most popular compatibilist theories of responsibility is described as a deep-self theory or sometimes as a mesh theory. This kind of theory comes in various flavors, but they all descend from Harry Frankfurt's higher-order desire theory (1988). Frankfurt distinguishes unwilling addicts (who desire not to desire drugs) from willing addicts (who desire to desire drugs or at least do not desire not to desire drugs). Frankfurt claims that unwilling addicts are not responsible, whereas willing addicts are responsible, because one is responsible when one's second-order desires "mesh" with one's first-order desires—when one desires to have the desires that one has.

To assess this claim about willing addicts, we need to distinguish several ways in which first-order and second-order desires can mesh. Some willing addicts might be willing because they enjoy the pleasure of the drug use and feel no desire to quit. Other willing addicts might not enjoy the use but still not have anything better to do and, hence, be willing *faute de mieux*. Still other willing addicts might be willing in the sense that they are not trying to quit, but the reason they're not trying to quit is precisely because they're addicted, so they have given up desiring to do what they know they cannot do. And yet more willing addicts might be willing only because they need drugs in order to avoid intense, chronic pain. Each of these kinds of willing addict could lead us to make different judgments about responsibility. So, if we want to understand responsibility, it's not enough to say that a person is a willing addict without saying more about why they are willing.

The most important kind of willing addict for our purposes are those who use drugs in order to cope with terrible life prospects, which Jeanette Kennett (2014) calls "resigned addicts." For example, imagine a hopeless homeless person living on cold, dangerous streets with no possibility of getting any safe housing or any menial job in the foreseeable future. He has no practical option other than life on the streets either with or without drugs. Life on the streets without drugs is so horrible for him that it is rational for him to want to use drugs, even if that means he remains addicted.

Are such resigned addicts fully responsible? We think not, and most people seem to agree, at least if the resigned addict is not at any fault for being homeless and hopeless. They might have some minimal degree of responsibility, but they are still not fully responsible, since they are not as responsible as the willing addict who is happy to use drugs just for pleasure. Other willing addicts, such as those who use drugs only to avoid intense, chronic pain, also do not seem fully responsible. These cases among others cast serious doubt on the claim that mesh between orders of desires is sufficient for full responsibility.

Scrupulosity poses a related problem. Some people with Scrupulosity are *ego-dystonic*, which means that they want not to be so scrupulous. They want not to want to be so limited by their moral concerns that they're unable to do other things they wish they could. They are more like unwilling addicts.

Other people with Scrupulosity, however, are *ego-syntonic*, which means that they do not object to their moral concerns, however extreme. They might wish that morality did not require so much of them, but they still might think their current life is the morally best life and the overall best life despite personal sacrifices. They view their own distress and anxiety as the cost of being moral, so they do not oppose it anymore than does someone who is willing to pay a cost in order to keep a promise, help the needy, or perform some other morally required action. They are resigned to their personal losses, because they accept perfectionist moral requirements and see no morally permissible option. In this respect, they are like resigned addicts who are resigned to addiction while homeless or chronic pain sufferers who are resigned to a life of addiction to pain-killers, because they see the other option as worse. They are not like willing addicts who use drugs just for pleasure, so they do not seem fully responsible.

These cases of ego-syntonic Scrupulosity create trouble for deep-self and mesh theories of responsibility. Those theories of responsibility imply that such agents are fully responsible because they endorse—or at least do not oppose—their Scrupulous impulses. On this view, if Adam is ego-syntonic, then he would be fully responsible for hurting the feelings of his children when he leaves them to check his receipts one more time. The problem is that Adam, much like a resigned addict or the addicted chronic pain sufferer, does not seem to be fully responsible. That is why we can feel pity for him rather than, or at least in addition to, anger. Thus, people with Scrupulosity, like resigned addicts, challenge Frankfurt's version of the deep-self or mesh theory of responsibility.

A related but distinct version of a deep-self theory is developed by Chandra Sripada (2016). Instead of referring to second-order desires, Sripada identifies a person's deep self with cares: "a person's deep self consists of her *cares*" (2016, 1206). The trick, of course, is to define cares. According to Sripada, a person cares about something only if four conditions are met:

(a) the person has "*intrinsic* motivation" to seek it,
(b) the person is motivated "to continue caring" about it,
(c) the person forms "normatively favorable" judgments of it, and
(d) the person feels "positively valenced emotions" about it. (2016: 1209–10)

These cares are expressed in actions: "An action expresses one's [deep] self if and only if the motive expressed in the action is one of one's cares" (2016: 1216). The inference to responsibility is then simple: "A person is morally responsible for an action if and only if it expresses her deep self" (2016: 1205).

This theory is very insightful and illuminating about attributability, but here the question is whether it captures accountability, specifically with regard to distinctive acts of Scrupulosity. Patients with ego-syntonic Scrupulosity do seem to have cares that (a) are intrinsically motivating, (b) they want to continue caring about, (c) they judge to be good, and (d) they feel positive emotions about. For example, assuming that Adam is ego-syntonic, Adam cares about checking his receipts in order to make sure that he pays the store all that he owes. He is intrinsically motivated to be sure about this because he seeks certainty not only in order to relieve his anxiety but also because he views it as his duty to be sure that he is not failing to pay anything he owes. His repeated acts of checking receipts from every store show that he wants to continue being sure that he pays all that he owes. He judges that it is good to be sure that he pays all that he owes. And he feels positive emotions about paying all that he owes as well as about being sure that he pays all that he owes, because he reports beliefs that it is morally required. Thus, Adam has cares about being sure that he pays all that he owes, according to Sripada's definition of cares. Adam is also motivated by these cares when he checks his receipts, so these actions express his cares and his deep self. Hence, Adam is morally responsible for these actions, and he knows that they cause the hurt feelings of his children. According to Sripada's theory, therefore, Adam is fully responsible for his actions and their harmful consequences.

The problem, as with Frankfurt's theory, is that Adam does not seem to be fully responsible for his actions or their consequences, as we argued above, and there's no obvious way to explain why an ego-syntonic Adam with Scrupulosity is any less responsible than an Adam without Scrupulosity who also ignores his children's needs because he prefers to watch a football game. We don't feel any pity for football-watching Adam at least in otherwise normal circumstances, but we do feel pity for Scrupulosity Adam. If we feel anger towards either of them, it is more towards the one who neglects his children to watch football. This reduction in anger, assuming the anger is justified, is a sign (even if not conclusive proof) that Adam is not fully responsible (Shoemaker 2015).

So we are left distinguishing the Adam with Scrupulosity from, on the one hand, the Adam who neglects his children to watch football and, on the other hand, the Adam who doesn't help his children with their homework because he has been kidnapped. Adam with Scrupulosity is partly, but only partly, responsible for his actions and their consequences, less than football-watching Adam and more than kidnapped Adam. However, Sripada's theory implies that Adam with ego-syntonic Scrupulosity is fully responsible. So much the worse for Sripada's theory.

We leave it to others to determine how Sripada could or should respond to this challenge. He could embrace this counterintuitive conclusion, distinguish ego-syntonic Scrupulosity from deep-self concerns, or find some way to explain away our intuitions. Each of these replies strikes us as problematic.

There are, of course, other versions of deep-self theories of responsibility, but they will run into many of the same problems regarding ego-syntonic Scrupulosity as well as resigned addicts. Instead of multiplying the variations on deep-self theories or countering cases and intuitions with more cases and intuitions, a more fruitful approach is to provide an alternative view and show how illuminating it is for Scrupulosity. So that is what we will do.

6.7 Reasons-Responsiveness Theories of Moral Responsibility

A competing theory of responsibility can explain why people with Scrupulosity are partly but not fully responsible. This popular theory understands an agent's responsibility for an action in terms of the agent's ability to respond to reasons for and against the action (cf. Fischer 1998). More specifically,

- An agent is *responsive* to reasons regarding a particular kind of act if and only if that agent is both reactive and receptive to reasons regarding that particular kind of act.
- An agent is *receptive* to reasons regarding a kind of act if and only if both
  - (p) If there is a reason for the agent to perform an act of that kind, then the agent usually will recognize that reason to perform an act of that kind.
  - (n) If there is a reason for the agent *not* to perform an act of that kind, then the agent usually will recognize that reason *not* to perform an act of that kind.
- An agent is *reactive* to reasons regarding a kind of act if and only if both
  - (p) If the agent recognizes overall reason to perform an act of that kind, then the agent usually will perform an act of that kind, and
  - (n) If the agent recognizes overall reason *not* to perform an act of that kind, then the agent usually will *not* perform an act of that kind.

What reduces moral responsibility is not simply failure to respond to reasons but rather inability to be responsive to reasons. That capacity is supposed to be lacking in addicts and others when they lack moral responsibility.

Our question is whether agents with Scrupulosity have reduced responsibility, according to this theory. The crucial point in this case is that those with Scrupulosity, at least in extreme cases, suffer from intense anxiety. Just as severe pain can make a person unable to think about anything else, so intense anxiety can also make a person unable to respond to any other reasons. As a result, people with extreme Scrupulosity display several features that show their lack of responsiveness to reasons.

First, in ways we canvassed above, people with extreme Scrupulosity tend to tailor their moral beliefs and actions to what soothes their anxiety instead of to the real reasons for and against actions, including moral rules and the welfare of others. They believe that what they did was terribly wrong when that belief helps them make sense of why they feel so guilty or so anxious, even if their guilt or anxiety is disproportionate to how much harm they caused to others. If they had caused less (or more) harm, they would believe that their act was just as wrong, and they would feel just as bad about it. Second, individuals with extreme Scrupulosity typically exaggerate by responding in major ways to minor infractions. What others see as insignificant foibles, they see as deadly sins, which magnifies their anxiety and guilt feelings. Third, those with extreme Scrupulosity tend to fixate on a subset of reasons. They focus on only one moral feature—or only a small set of moral features—that they have found relieves their anxiety. This fixation makes them unreceptive to other moral reasons.

All of these failures result from the kind of underlying anxiety that guides and shapes their moral beliefs and actions. Because their anxiety is so persistent and intense, at least in extreme cases, their mental illness makes them unable to respond appropriately to reasons. That is why their responsibility is reduced. On the other hand, they remain able to respond to some reasons. They are not always in the grip of intense anxiety, and they are usually able to respond to extremely strong reasons. Adam would stop checking receipts if his fire alarm went off and he needed to save the lives of his children. Linda would take a shower if her doctor told her that she had just been exposed to a deadly infection. Reasons-responsiveness theories can, thus, explain why people who act out of Scrupulosity have some responsibility, but their responsibility is reduced.

Some reasons-responsiveness theorists (such as Fischer 1998) explicitly deny that responsibility comes in degrees. However, this claim is neither plausible nor necessary for the theory, since other reasons-responsiveness theorists can accept degrees of responsibility (Coates and Swenson 2013; Nelkin 2016). Indeed, it should come as no surprise that responsiveness to reasons comes in degrees. People respond more or less consistently to more or fewer reasons. That admission of degrees of responsibility is part of why this theory is so well-suited to account for intermediate cases like Scrupulosity.

Such theories can also explain degrees of responsibility in other cases. Agents can be more or less able to respond to more or fewer reasons in more or fewer situations. These variations can explain how an addict (or an agoraphobe) who neglects his children can be more responsible than if he were kidnapped but less responsible than if he neglected his children solely because he did not like them. Each of these mental illnesses comes in degrees that vary with degrees of responsiveness to reasons and, hence, also vary in degrees of responsibility. The fact that reasons-responsiveness tracks responsibility in such a wide variety of cases makes such theories useful and attractive.

6.8 Conclusions

We have argued that people with ego-syntonic as well as ego-dystonic Scrupulosity are partly but not fully accountable or responsible because they are partly but not fully responsive to moral and personal reasons. We need to say much more in order to fully explain, support, and defend these conclusions. For example, some people with Scrupulosity might seem to be fully responsible for their present actions because they were fully responsible for their past actions that caused them to have Scrupulosity, as suggested by tracing principles (see Vargas 2005). We doubt that any such considerations will undermine our main conclusions, but showing that is a topic for elsewhere (especially Summers and Sinnott-Armstrong 2019).

References

American Psychiatric Association. 2013. *Diagnostic and Statistical Manual of Mental Disorders: DSM-5.* Washington, DC: American Psychiatric Association.

Ciarrocchi, J. 1995. *The Doubting Disease: Help for Scrupulosity and Religious Compulsions.* Marwah, NJ: Paulist Press.

Coates, D. J., & Swenson, P. 2013. Reasons-responsiveness and degrees of responsibility. *Philosophical Studies* 165 (2): 629–45.

Fischer, J. M. 1998. *Responsibility and Control.* Cambridge: Cambridge University Press.

Frankfurt, H. 1988. Freedom of the will and the concept of a person. In *The Importance of What We Care About* (pp. 11–25). Cambridge: Cambridge University Press.

Kane, R. 2002. *The Oxford Handbook of Free Will.* New York: Oxford University Press.

Kennett, J. 2014. Just say no? Addiction and the elements of self-control. In N. Levy (ed.), *Addiction and Self-Control.* Oxford: Oxford University Press.

Kurth, C. 2018. *The Anxious Mind.* Cambridge, Mass.: MIT Press.

Nelkin, D. K. 2016. Difficulty and degrees of moral praiseworthiness and blameworthiness. *Noûs* 50 (2): 356–78.

Pereboom, D. 2001. *Living without Free Will.* Cambridge: Cambridge University Press.

Santa, T. M. 2007. *Understanding Scrupulosity: Questions, Helps, and Encouragement.* Ligouri, MO: Liguori/Triumph Publishers.

Shoemaker, D. 2011. Attributability, answerability, and accountability: Toward a wider theory of responsibility. *Ethics* 121: 602–32.

Shoemaker, D. 2015. *Responsibility from the Margins.* Toronto: Oxford University Press.

Singh, D. and Sinnott-Armstrong, W. 2015. The DSM-5 definition of mental disorder. *Public Affairs Quarterly* 29 (1): 5–31.

Sripada, Chandra. 2016. Self-expression: a deep self theory of moral responsibility. *Philosophical Studies* 173: 1203–32.

Summers, J. and Sinnott-Armstrong, W. 2019. *Clean Hands: Philosophical Lessons from Scrupulosity.* New York: Oxford University Press.

Vargas, M. 2005. The trouble with tracing. *Midwest Studies in Philosophy* 29 (1): 269–91.

Watson, G. 2004. *Agency and Answerability: Selected Essays.* Oxford: Clarendon Press.

7

Addiction and Agency

Justin Clarke-Doane and Kathryn Tabb

7.1 Introduction

It is often thought that there are certain sorts of causal factors that should mitigate attributions of blame or praise. Certain psychological processes that lead people to act, for example, may be thought to render typical punishments or rewards unfair, and require a different sort of moral response. A paradigmatic case is that of addiction, insofar as addicts are often seen as lacking full freedom resulting from their compulsive prioritization of using over all else. Often in the philosophical literature, as well as in popular media, the character of the addict is portrayed as compelled or "seduced" by their addiction (Cummins 2014; Grim 2007, 191). For example, Gorski and Miller begin their text:

> Addiction is distinguished from [mere heavy] drug use by the lack of freedom of choice. Using a mood-altering substance is a choice. Addiction is a condition that robs a person of choice and dictates the frequency, the quantity, and the nature of use (1986, 39).[1]

The paradigm of addiction is therefore useful for philosophers interested in thinking about free will and moral responsibility, but worried about the possible scope of mitigating causal histories (Berofsky 2005; Kane 2020; Levy 2011; Shatz 1988; Yaffe 2011). The compulsive prioritization of using over other goals is taken to indicate a difference in kind between the addict and the rest of us, and gives grounds for delineating exceptional cases from typical ones when it comes to assigning desert.

---

[1] See Nadelhoffer 2010 and Pickard 2017 for perspectives on this narrative that are consonant with the argument to follow.

Justin Clarke-Doane and Kathryn Tabb, *Addiction and Agency* In: *Agency in Mental Disorder: Philosophical Dimensions*. Edited by: Matt King & Joshua May, Oxford University Press. © Justin Clarke-Doane and Kathryn Tabb 2022. DOI: 10.1093/oso/9780198868811.003.0008

Helping ourselves to paradigmatic cases like addiction can give the illusion of progress in debates over moral responsibility. Facing a challenge over the mitigating potential of causal histories, for example, the believer in moral responsibility might claim that an individual is culpable when their action is intentional, counterfactually dependent on their intention, and not motivated by whatever kind of causes lead addicts to act as they do. Such an argument makes use of what we can call the *method of paradigms,* where a case about which intuitions are supposedly clear is used to guide reasoning about other cases. This method is prominent in the literature on moral responsibility for an obvious reason: it bypasses the problem of specifying what sort of causal history should be taken as exculpatory by ostension, through specifying a condition that is generally thought capable of mitigating moral responsibility and generalizing from there.

Let's consider three examples of addictive behavior. On the "folk philosophical view" each of these behaviors would count as morally mitigated (even if not exculpated).

Example 1: The cocaine addict who steals someone's TV to buy more cocaine.

Example 2: The gambling addict who gambles away his children's college fund.

Example 3: The sex addict who commits adultery.

Now consider analogs to (1)–(3). On the folk philosophical account, these would *not* count as morally mitigated.

Example 1*: The non-cocaine-addict who steals someone's TV because he wants to indulge his habit of watching TV over dinner, but his own TV set is broken.

Example 2*: The first-time gambler who gambles away his children's college fund because of careless probability judgments during a business trip to Las Vegas.

Example 3*: The non-sex-addict who commits adultery to fulfill his self-image as a pickup artist.

How can we distinguish between (1)–(3) and (1)*–(3)* with respect to moral responsibility? The method of paradigms presumes that there is no way to fill in the details so that (1) and (1)*, (2) and (2)*, or (3) and (3)* both satisfy, or fail to satisfy, the conditions for responsible action. But our question is this: on what grounds can we conclude that the circumstances of (1)–(3) are mitigating, while those of (1)*–(3)* are not, *independent of the assumption that the non-addicts in (1)*–(3)* differ from the addicts in (1)–(3) by being responsible?* Without an informative account of when a person's causal history is sufficiently like that of an addict, the method of paradigms seems to provide only question-begging grounds for making such determinations.

By appealing to a biomedical concept, it might seem one can avoid the question of what makes an addict's actions different—on the assumption that, since clinicians and researchers seem to know what addiction is, it must represent a distinct class, with underlying properties that can explain the unique ways in which addicts act. We will argue that, on the contrary, our best scientific theories of addiction suggest that its essential features can be found in other processes that are *not* intuitively mitigating. We discuss four prominent models of addiction, and show that all of them explain addiction in terms of psychological processes that in other contexts are *not* supposed to diminish one's responsibility. The upshot is that if addiction mitigates on account of these features, then so too do other conditions that seem irrelevant for questions of responsibility.

While our arguments do not pose a problem for clinicians or researchers (or philosophers) aiming to understand addiction, they do pose a problem for moral philosophers using the addict as a paradigmatic case. If similar problems plague appeals to other characters familiar from the responsibility literature, like "obsessive compulsives," "sociopaths," and Tourette Syndrome patients—as we expect they do—then our conclusions are of general significance.[2] Actions resulting from conditioned learning, faulty reasoning, unfortunate, undesirable, or unusual identities, or various personal histories, may all be like actions resulting from addiction. We conclude that, by so expanding the boundaries of moral responsibility, the method of paradigms tends to support a more general skepticism about moral responsibility.

7.2 What Is an Addict?

We should start by noting that there is hardly an agreed-upon folk definition of addiction, much less a clinical one. The heroin user, the alcohol abuser, the sex addict, and so forth are often taken to *decide* to use, drink, or cheat,

---

[2] We use these designations in quotation marks to indicate not only our suspicion that these labels do not refer to natural kinds but also to indicate our discomfort with arguments that make instrumental use of such caricatures.

but in a way that is in an important sense different from that of other intentional beings. Terms like "compulsive," "irresistible need," "persistent dependence," and "loss of control" are employed to capture this difference; but there is little consensus among laypeople or experts about what they mean. After an extensive review of working definitions of addiction, a report commissioned by the European Monitoring Centre for Drugs and Drug Addiction concluded that "accumulated evidence indicates that impaired control, conflict, craving and so on are not necessary features of addiction even though they are frequently observed and have to be accounted for in any comprehensive theory" (West 2013). Addicts can resist using for days or weeks on end when sufficiently motivated (Hart et al. 2000) and, as the report explains, common symptoms like craving, withdrawal, and increased tolerance are not universal features. Accordingly, its author, Robert West, defines addiction as "a repeated powerful motivation to engage in a purposeful behaviour that has no survival value, acquired as a result of engaging in that behaviour, with significant potential for unintended harm" (27).

While West resists specifying putative mechanisms that might more narrowly define "powerful motivation," or even committing to the existence of such mechanisms, he acknowledges that addiction is widely viewed as a categorically distinct pathological state, even by constituencies who agree on little else. For example, advocates of the biomedical model have increasingly spoken of addiction as a "brain disease" (Leshner 1997; Volkow et al. 2016) since the *Diagnostic and Statistical Manual of Mental Disorders* first recognized it as a primary mental health disorder rather than just a symptom of underlying psychopathology (Robinson and Adinoff 2016). And while avoiding the biomedical model, addicts who adopt the tenets of Alcoholics Anonymous also defend the view that addiction is a disease, contrasting it with everyday cases of weakness of will and sinfulness.

The stakes of this question are high, as proclaimed in the title of Leshner's "Addiction is a Brain Disease, and It Matters." Here Leshner argues, "The gulf in implications between the 'bad person' view and the 'chronic illness sufferer' view is tremendous" (45). Like many advocates of the biomedical approach, he believes that seeing addiction as a disease of the brain will reduce social stigma and transform the way the addicts are treated by the public health and criminal justice systems. He suggests that recognizing that "an addict's brain is different from a nonaddict's brain" (46) could allow us to distinguish those with a disease from "weak or bad people, unwilling to lead moral lives and to control their behavior or gratification" (45). Additionally, "Elucidation of the biology underlying the metaphorical switch is key to the development of more effective treatments, particularly antiaddiction medications" (46).

However, the demarcation criteria for what counts as a disease—that is, what would flip Leschner's "switch"—are, as philosophers of medicine have long pointed out, disturbingly unclear (Bingham and Banner 2014; Stein et al. 2010). Indeed, the lack of a widely accepted conceptual analysis of the category of disease means that there is no easy answer as to whether addiction is or is not a disease, nor accordingly whether any individual who uses a substance has a disease or not. *Pace* Leshner, the problem is not solved by the fact that subjectively identified signs and symptoms have been discovered to have neurological correlates in the reward system, affective systems, or executive control system of the brain (which they have; for an overview see Volkow and Boyle 2018). Variation is not pathology; understanding the mechanics of blood pressure is distinct from inventing a criterion for hypertension, as the latter results from taking a stand on what counts as too high a reading. Likewise, the discovery of recognizable mechanisms underlying impaired motivation does not answer the question of when motivation should be considered impaired. This problem will endure until a consensus forms around how *much* users need differ from non-users (who also, from time to time, display suboptimal functioning with respect to motivation, inhibition regulation, and compulsivity) before they are called an addict.

However, for our purposes, what matters is not whether addiction is a disease per se, but whether it is a mitigating condition. Not all diseases—even all mental diseases—are mitigating. And some mental states are mitigating without being diseases. In the courtroom, for example, a diagnosis is less relevant for assessing culpability and for sentencing than the displaying of certain features, such as a floridly psychotic state at the time of the crime or cognitive deficit like dementia or low IQ. From this perspective, addicts are exculpated not *for being addicts*, but rather for meeting some other measure. In that case, what would matter is whether non-addicts periodically meet that measure, too, rather than the extent to which they resemble addicts in other ways. Even if there were universal agreement about who counts as an addict, then, it would not solve our worries about the method of paradigms, without an accompanying theory of who is *like* an addict in the sense relevant for assessing desert.

7.3 Theories of Addiction

What are the most prominent scientific theories of addiction? In this section we will examine four, each of which would be a natural place for the philosopher to turn when looking to identify the mechanism whose presence can justify judging addicts differently. Ideally, one or more of these accounts would provide an explanation for addiction that could work in tandem with our best accounts of responsibility to explain the common intuition that addicts are less responsible for their actions than non-addicts. That explanation could guide other judgments about mitigation, ideally validating intuitions about what sort of histories matter for desert. If one or more account of addiction could do this work, we would be able to save the method of paradigms from an unpleasant dilemma: on the one hand, falling into a vicious circularity (anyone who is importantly like an addict has mitigated responsibility, and to be importantly like an addict is to have mitigated responsibility in the way as an addict does) or on the other, accidently excusing more of the actions of non-addicts than many would be comfortable with.

First worth considering are those accounts of addiction that employ the terms of operant learning theory. For example, one influential model, the incentive-sensitization theory, posits that addicted behavior is caused by an increased sensitivity in the brain to the reward-value of certain substances or behaviors, such that the brain learns to "want" drugs even if they are not (or are no longer) "liked." Even if the opiate user no longer feels euphoria when taking a hit, they may still feel an intense craving when they see drug paraphernalia that will lead them to desire to use. Evidence for this account is drawn from animal models showing how reward cues are mediated by dopamine-related systems in the addicted brain: "addicted" animals are those who "have stronger cue-triggered urges and intensely 'want' to take drugs [....] addiction becomes compulsive when mesolimbic systems become sensitized and hyperactive to the incentive motivational properties of drug cues" (Berridge and Robinson 2016, 673). Berridge and Robinson concede that their theory does not demarcate the addict from the non-addict, but maintain that "incentive sensitization can make the temptations faced by addicts harder to resist than those most other people are called upon to face" (675). This justifies, in their view, calling addiction a "brain disease" (675).

A second approach to explaining why addicts struggle with motivation to abstain focuses on reflective choice. One example of such is the hyperbolic discounting model advocated by George Ainslie. In Ainslie's picture, behavior that seems compulsive is not due to a weakness of the will or a failure of the individual to make a certain choice, but to the outcome of what Ainslie refers to as an intrapersonal "marketplace of reward," in which different interests compete across timescales, producing effects like hyperbolic discounting, in which far-off goals lose their motivational power in favor of immediate gains. With respect to demarcating addiction, Ainslie is comfortable with the notion that his account might be revisionist: "If addiction is defined with a low threshold, half the people in America are addicted to something [....] Those of us who have avoided the named addictive diagnoses are nevertheless apt to suffer from habitual overvaluation of the present moment, as in chronic procrastination, overuse of credit, or unrealistic future time commitment" (Ainslie 2018, 37). The question to answer about those we consider addicts is not what makes them struggle with cognitive effects like hyperbolic discounting, but what makes these problems so extreme for them, and what stops them from using the usual methods that non-addicts use to conform to social expectations around choosing. In other words, if addiction is a disease at all, it is "a disease of motivation, that is, one that does not bypass the mechanism of choice" (42).

A third theory sees addiction as due to the malformation of the identity of the addict, either due to social pressures and cues from the environment, individual traumas, or positive influences on identity, like group membership (Walters 1996). Wasmuth et al. have conceived of addiction as an occupation, that is, a self-organizing human activity that provides "meaning, temporal structure, roles, habits, routines, and volition to individuals" (Wasmuth et al. 2014, 605). Understanding addictions in this way allows the authors to explain the failures of abstinence to constitute recovery, in so far as it may "be profoundly distressing because of [...] not having or being able to participate in occupations that were once central to daily living" (607). Alcoholics Anonymous, Narcotics Anonymous, Sex Addicts Anonymous, and so forth address this by allowing for the renegotiation of an addict's identity, first as an addict and then as an addict in recovery (Best et al. 2016). Philosophers have also discussed the importance of recognizing that the identity of being an addict can exert a substantial pull, and can be a response to environments in which other sources of identity are hard to acquire or maintain (Tekin et al. 2017). According to these sorts of theories, changing how people who struggle with addiction imagine their own self-efficacy and agency can be transformative (West 2013, 59).

Finally, there are models of addiction that attribute it not to factors within the individual, but factors within the environment. A provocative study of this sort was Alexander et al.'s so-called "Rat Park" experiment, which supported the hypothesis that addiction is more common in rats subjected to social deprivation and stimulus-poor environments (Alexander et al.

1978). In contrast, rats who lived in a stimulus-rich environment, where they were allowed to maintain their natural family structures, mutually groom, and play together, showed less interest in consuming addictive substances. In light of these experiments, Alexander and Hadaway argued that opiate addiction was better explained as a rational response to distress and deprivation than as a conditioned shift brought on by drug exposure (1982). More recently, Carl Hart has argued that the science behind claims that addiction is a brain disease is shoddy and misleading, and that addiction is better understood as the result of the psychiatric disorders often comorbid with it, and of socioeconomic factors like poverty, systemic racism, and unemployment (Hart 2017). According to Hart, there is nothing unique about the addicted brain that isn't also true of brains that undergo other sorts of stress and trauma brought about by similar circumstances.

7.4 The Problem with Paradigms

Although the theories of addiction surveyed in the previous section are not exhaustive, they give a good indication of the kinds of accounts today's best science suggest. What is striking is that none seems able to solve the problem of explaining why addiction would be mitigating in a way that other causal histories would not. Instead, theories of addiction tend to provide a scale of function on which addicts are taken to cluster at the low end. None of these scales of function correspond in any straightforward way with our intuitions about moral responsibility. So, without a clear indication of where on the scale one becomes "like an addict," they provide at best an imprecise guide. If one gives up on that sort of specificity, and rules that anyone who acts in a way that is abnormal with respect to one of these functions is suitably like an addict, one would end up exculpating a whole host of everyday figures like the hot-head, the hedonist, the egoist and the victim of circumstance—that is, all of us, in our less proud moments. In this section we consider why this is the case.

Philosophical theories of freedom and responsibility vary widely, but most locate agency at least partly in psychological processes that allow one to respond to reasons that express one's "deep self" or true values. Young children and squirrels, for instance, are not morally responsible for stealing a sandwich because they are simply incapable of controlling or guiding their actions in light of reasons related to property rights. Similarly, typically adults are less blameworthy for an insensitive remark if it was out of character, is something that they wholly disavow, or otherwise fails to "mesh" with their higher-order convictions. The trouble is that, regardless of which particular philosophical approach one takes toward freedom and responsibility, none yield a categorical difference between addictive and non-addictive choice, in tandem with the science.

Let us begin with the idea that, e.g., alcohol abuse is mitigating because "alcoholism is a brain disease." What plausible theory of responsibility could deliver this verdict? By calling something a brain disease we mean that there are known neurobiological correlates for a recognized category of disorder. The existence of such neural correlates in addiction need not impact the user's desires such that they fail to mesh with their higher-order convictions, in the sense of Frankfurt (2003 [1971]). Frankfurt and followers argue that people are not responsible for their actions if the desires that led to those actions are in conflict with more deep-seated features of their psychology. Nor do neural correlates preclude the alcoholic's "valuing" his addiction in the sense of Watson (1987). Watson suggests that an action is unfree when the agent's pursuits are not in alignment with what they value due to internal dysfunction; we are all familiar with figures like Keith Richards or William Burroughs who thoroughly valued their addiction, though most people sympathetic to the biomedical view of addiction would say they had a disease. Did their using at least undercut their capacity for "guidance control" in the sense of Fischer and Ravizza? They argue that it is not the capacity to do otherwise per se, but the capacity to do otherwise in response to reasons that matters for moral responsibility (2000). Again, it does not seem so. One can have a brain disease while all of the requisite rational capacities remain intact (as with, say, chronic migraines). Nor need a brain disease disrupt the "sane deep self" in the sense of Wolf (2012); the simple fact of cognitive pathology does little to support the normative conclusion that the subject's deep self is not functioning *well*. For example, many disorders described in the DSM do not bring about the ipseity disturbance typical of psychosis; think of the general anxiety disorder clinicians often use to diagnose the "worried well." Evidently, whether having a brain disease mitigates moral responsibility depends on whether the brain disease engenders compulsion, and so the biomedical definition of brain disease is, by itself, insufficient to explain why addicts are less responsible than non-addicts.

It might be thought that learning models like the incentive-sensitization model hold more promise insofar as they suggest that, for some of us, substances or activities become "wanted" in a way that is out of proportion with how much we "like" them. In order to think about them in terms of a mesh theory like Frankfurt's, one might reframe this model in terms of the substances or activities being desired to a degree that fails to align with reflective preferences. The problem is that the constant, unreflective revaluing of stimuli in our environment on the basis of dopaminergic rewards, made possible by the extreme plasticity of neurodevelopment, means that "wanting" fluctuates with respect to *everything* we perceive to be of interest to us—not just drugs and addictive behaviors. As Lewis writes, "When the brains of addicts (following years of drug taking) are compared to those of drug-naive controls, these scientists can be heard to say 'Look! Their brains have changed!' Yet if neuroplasticity is the rule, not the exception, then they're actually not saying much at all. The brain is supposed to change with new experiences" (2017, 10).

"Wanting" more than we "like" is, as Lewis notes, a common experience outside of addiction. Think of the college student who binges on Netflix instead of studying, the Shakespearean heroine who follows her beloved despite being scorned and abused, or the athlete who is so set on competing that she pushes her body beyond the limits of what she enjoys. The fact that our folk category of addiction is often stretched to accommodate cases like these (think of Robert Palmer's "Addicted to Love") means that the question of what sort of conflicts between "wanting" and "liking" counts as *addictive* is really just a question of ethics. On the biological level, the change in activation from the ventral to the dorsal striatum associated with compulsion has been demonstrated to occur in many other circumstances, including falling in love (Lewis 2017). So while a mesh theory of moral responsibility might supply terms for taking the addict's actions as a paradigmatic case of mitigated blame in our moral reasoning, the result will be that many of our everyday choices will no longer seem to accord with our higher-order desires either, and our responsibility for them will be mitigated too (Pickard 2015, 2017).

It might be thought that rational choice accounts of addiction, in conjunction with some version of the guidance control theory, constitute a more promising approach. Moral agents act on the basis of reasons—that is, as a result of psychological mechanisms that are reason-responsive and which play a causal role in their choices to act—even if they could not have acted otherwise. So, attributing to addiction a pathological method of choosing might seem to qualify the addict for exemption from moral responsibility, according to this theory. But as noted above, the leading accounts of addiction that theorize it as a pathology of choice do *not* say addicts choose in a way different in kind from the rest of us. They are simply at an extreme of functioning with respect to certain kinds of universal decision-making. Hyperbolic discounting, paradigmatic of addicts according to Ainslie, is ubiquitous in children, and undeniably frequent in adulthood too. It can also vary within the same individual over time: just like other sorts of moral reasoning can be swayed by circumstances (Ditto et al. 2009), hyperbolic discounting can be improved when people are assisted in contemplating their future selves in a concerted way before choosing (Hershfield et al. 2011). If one wishes to say that addicts are exculpated from moral responsibility because they are not responding in the right way to reasons and thus are not exerting the right kind of guidance control, appearances suggest that one would need to say the same of all of us in those cases where, for example, we skip the gym despite our best intentions.

We have still not considered identity-based explanations of addiction, which are congruent with valuing theories of responsibility. These take moral responsibility to turn on an action's compatibility with the agent's "deep self". Wasmuth et al.'s success at replacing one occupation (addiction) with another (theater)—as well as the impressive efficacy of AA's encouragement of a transformative social role—suggest that addiction affects addicts like other occupations, shaping "not only their surroundings but also their personal identities, values, and personal roles" (607). To this extent, then, it would be question-begging to declare that addiction is mitigating, rather than the result of individuals acting in accordance with their deeply-held values. One would need to show that, contrary to substantial evidence, addiction is not an identity, or not the right kind of identity to engender personal values.

This worry is in the spirit of Wolf's criticisms of valuing theories (2012). She argues that they fail to attend to the source of our values. If the true self originates in trauma or other corrosive factors, the individual may be incapable of forming the right values, and thus not responsible for their true self. The problem is that the causal history of the addict need not be uniquely traumatic, nor uniquely anything. Many people with unfathomably difficult personal histories do not become addicts, and many lacking that sort of precipitating cause do. So, while one might want to say that identities *like* addiction are mitigating, this just returns us to the question of what it means to resemble an addict in a way that matters. It is hard to see how we could, in a principled way, delineate between people who resemble addicts but whose behavior fails to count as mitigated—like that of the hypercompetitive stock-trader or the besotted lover—from those people that are excused on the grounds that they just *can't* make the right choice.

7.5 Taking Stock

Is there another pair of theories—of addiction on the one hand and responsibility on the other—that might justify our intuitions about the mitigating power of addiction? Not that we are aware of. We suggest that, with the paradigm of the addict, the method of paradigms does not elucidate much. Science cannot save the day by opening the black box of addiction and explaining why it would be peculiarly mitigating, compared to the circumstances in which non-addicts regularly find themselves. All science can do is tell us whether addicts are categorically different from non-addicts, and it seems to have told us that they are not.

If addiction is conceived of as mitigating because, e.g., it is the result of socioeconomic hardship, it is hard to see how we could avoid the conclusion that all manner of antisocial actions should be excused when they too result from difficult personal history. When Hart describes, in his memoir *High Price* (2013), the conditions which make the dealing and consuming of drugs a rational choice, he describes the intentional disempowering and oppression of minority communities. Insofar as addiction is a proximate cause of, say, a criminal action and conditions like these are the ultimate cause, it is arbitrary to regard addiction as what is mitigating, rather than the conditions themselves. Indeed, a reasonable next step is to ask why we don't view an environment of deprivation and injustice as itself a mitigating circumstance, whether or not addiction plays a mediating role. The accounts of addiction we considered above suggest other widespread features of being human that, if similarly accepted as mitigating, would liberalize our notions about desert.

Of course, it remains possible that there are other paradigms, such as "obsessive-compulsives," "sociopaths," or Tourette Disease patients from which we might abstract instead, which would more narrowly demarcate the kind of causal histories that should be seen as diminishing blameworthiness. Maybe the problem is with addiction, not with the method of paradigms. But we submit that any such paradigm drawn from psychopathology is likely to engender analogous problems. For example, obsessive-compulsives can also have desires that mesh, can value their behaviors, can be reason-responsive by any ordinary standard, and may exhibit guidance control (see Summers & Sinnott-Armstrong, this volume). So, it does not help much to say that someone is less responsible for their behavior to the extent that it is like the obsessive-compulsive's. *What is it* about the obsessive-compulsive's behavior that could be mitigating? Absent an answer to this question, such slogans are without clear content.

Within psychiatry, concerns have been expressed about the reification of psychiatric diagnoses, which can increase stigma and provide obstacles to the biomedical exploration of the full range of psychopathology (Hyman 2010; Tabb 2017). Another casualty may be in philosophy, where the use of diagnostic kinds as paradigmatic cases risks obscuring the complexity of outstanding problems in moral philosophy. But it may be that the paradigm of the addict can help clarify our intuitions in a different way, by bringing into view a radical dilemma. If non-addicts display the key features of addiction to some degree or other, and if the line can only be drawn contingently and subjectively between the normal and the pathological, it would seem that we either need to stop excusing addicts, or start excusing others. We suggest that a progressive step would be to consider what feels fair about mitigating our judgments of blameworthiness when it comes to addicts, and to examine the barriers that stop us from extending the same dispensation to others who display intemperance from time to time—that is, to everybody.

References

Ainslie, G., 2018. The picoeconomics of addiction. In *The Routledge Handbook of Philosophy and Science of Addiction*, 54–64. New York: Routledge.

Alexander, B.K. and Hadaway, P.F., 1982. Opiate addiction: The case for an adaptive orientation. *Psychological Bulletin*, 92(2): 367.

Alexander, B.K., Coambs, R.B., and Hadaway, P.F., 1978. The effect of housing and gender on morphine self-administration in rats. *Psychopharmacology*, 58 (2): 175–9.

Berofsky, Bernard, 2005. Ifs, cans, and free will: The issues. *Oxford Handbook of Free Will*. Oxford: Oxford University Press.

Berridge, K.C. and Robinson, T.E., 2016. Liking, wanting, and the incentive-sensitization theory of addiction. *American Psychologist*, 71(8): 670.

Best, D., Beckwith, M., Haslam, C., Alexander Haslam, S., Jetten, J., Mawson, E., and Lubman, D.I., 2016. Overcoming alcohol and other drug addiction as a process of social identity transition: The social identity model of recovery (SIMOR). *Addiction Research & Theory*, 24(2): 111–23.

Bingham, R. and Banner, N., 2014. The definition of mental disorder: Evolving but dysfunctional? *Journal of Medical Ethics*, 40: 537–42.

Cummins, Denise. 2014. The myth and reality of free will: The case of addiction. How addiction robs us of free will, and how to outwit it. *Psychology Today*. Posted February 9, 2014. Available online at: <https://www.psychologytoday.com/us/blog/good-thinking/201402/the-myth-and-reality-free-will-the-case-addiction>

Ditto, P.H., Pizarro, D.A., and Tannenbaum, D., 2009. Motivated moral reasoning. *Psychology of Learning and Motivation, 50*: 307–8.

Fischer, J.M. and Ravizza, M., 2000. *Responsibility and Control: A Theory of Moral Responsibility*. Cambridge: Cambridge University Press.

Frankfurt, H., 2003 [1971]. Freedom of the will and the concept of a person. In G. Watson, ed. *Free Will*, 322–36. Oxford: Oxford University Press.

Gorski, T. and M. Miller. 1986. *Staying Sober: A Guide for Relapse Prevention*. Independence, MO: Herald House/Independence Press.

Grim, Patrick. 2007. Free will in context: A contemporary philosophical perspective. *Behav. Sci. Law* 25: 183–201.

Hart, C.L., 2013. *High Price: A Neuroscientist's Journey of Self-Discovery That Challenges Everything You Know about Drugs and Society*. New York: HarperCollins Publishers.

Hart, C.L., 2017. Viewing addiction as a brain disease promotes social injustice. *Nature Human Behaviour, 1*(3): 1–1.

Hart, C.L., Haney, M., Foltin, R.W., and Fischman, M.W., 2000. Alternative reinforcers differentially modify cocaine self-administration by humans. *Behavioral Pharmacology, 11*(1): 87–91.

Hershfield. H. 2011. "Future self-continuity: how conceptions of the future self transform intertemporal choice. *Annals of the New York Academy of Sciences*, 1235(1): 30-43.

Hyman, Steven E. 2010. The diagnosis of mental disorders: The problem of reification. *Annual Review of Clinical* Psychology, 6. Annual Reviews: 155–79.

Kane, Robert, 2020. Reflections on free will, determinism, and indeterminism. *The Determinism and Freedom Philosophy Website*. Undated web article. Available at: <https://www.ucl.ac.uk/~uctytho/dfwVariousKane.html>

Leshner, A.I., 1997. Addiction is a brain disease, and it matters. *Science, 278* (5335): 45–7.

Levy, Neil, 2011. Addiction, responsibility, and ego depletion. *Addiction and Responsibility*. Cambridge: MIT Press.

Lewis, M., 2017. Addiction and the brain: Development, not disease. *Neuroethics, 10*(1): 7–18.

Nadelhoffer, Thomas. 2010. Does free will disappear because of addiction? *Flickers of Freedom (Blog).* Posted June 21, 2010. Available at: <https://philosophycommons.typepad.com/flickers_of_freedom/2010/06/does-free-will-disappear-because-of-addiction.html>

Pickard, Hanna, 2015. Psychopathology and the ability to do otherwise. *Philosophy and Phenomenological Research,* 90(1): 135–63.

Pickard, Hanna, 2017. Addiction. *The Routledge Companion to Free Will:* 454–67.

Robinson, S.M. and B. Adinoff., 2016. The classification of substance use disorders: Historical, contextual, and conceptual considerations. *Behavioral Sciences,* 6(18).

Shatz, David, 1988. Compatibilism, values, and "could have done otherwise." *Philosophical Topics,* 16(1): 151–200.

Stein, D.J., Phillips, K.A., Bolton, D., Fulford, K.W.M., Sadler, J.Z., and Kendler, K.S., 2010. What is a mental/psychiatric disorder? From DSM-IV to DSM-V. *Psychological Medicine,* 40(11): 1759–65.

Tabb, K., 2017. Philosophy of psychiatry after diagnostic kinds. *Synthese,* 19: 2177–95.

Tekin, Ş., Flanagan O., and Graham G., 2017. Against the drug cure model: Addiction, identity, and pharmaceuticals. In: Ho, D., (ed.), *Philosophical Issues in Pharmaceutics. Philosophy and Medicine,* vol. 122. Dordrecht: Springer.

Volkow, N.D. and Boyle, M., 2018. Neuroscience of addiction: Relevance to prevention and treatment. *American Journal of Psychiatry,* 175(8): 729–40.

Volkow, N.D., Koob, G.F., and McLellan, A.T., 2016. Neurobiologic advances from the brain disease model of addiction. *New England Journal of Medicine,* 374: 363–71.

Walters, G.D., 1996. Addiction and identity: Exploring the possibility of a relationship. *Psychology and Addictive Behaviors,* 10(1): 9.

Wasmuth, S., Crabtree, J.L. and Scott, P.J., 2014. Exploring addiction-as-occupation. *British Journal of Occupational Therapy,* 77(12): 605–13.

Watson, G., 1987. Free action and free will. *Mind,* 96(382): 145–72.

West, R., 2013. *EMCDDA INSIGHTS: Models of Addiction.* Luxembourg: Publications Office of the European Union.

Wolf, S., 2012. Sanity and the metaphysics of responsibility. In Russ Shafer-Landau, ed., *Ethical Theory: An Anthology.* Oxford: John Wiley & Sons.

Yaffe, Gideon. 2011. Lowering the bar for addicts. In *Addiction and Responsibility.* Cambridge: MIT Press.

8

Mental Disorders Involve Limits on Control, Not Extreme Preferences

Chandra Sripada

8.1 Introduction

People with mental illness engage in characteristic disorder-associated behaviors. A person with obsessive compulsive disorder (OCD) washes their hands dozens or hundreds of times a day. A person with attention-deficit/hyperactivity disorder (ADHD) is distractible and disorganized and fails to complete their assigned tasks. A person with alcoholism drinks to excess, with resulting harms to work and family. How are we to make sense of why these people do what they do?

A standard position is that those with mental illness cannot help but do what they do. They have a disorder and what they do is not a matter of choice. We would not blame a person with acromegaly for having too much growth hormone; so too we should not blame a person with ADHD for distractedly forgetting to go to an appointment.

There are two major shortcomings of this simple "disease model" of mental illness. First, it seems to require two different models to explain action. Most purposive actions are explained in the usual way in terms of the ordinary workings of our motivational psychology—beliefs, desires, deliberation, etc. Some purposive actions, the disorder-associated actions of those with mental illness, get explained in a quite different way. For these special cases, a "disease-based" process is invoked, though the particulars of how this process works are not filled in with any detail. Splitting up explanations in this way, especially without providing details on how the second kind of explanation is supposed to work, seems ad hoc. Second, while the person with acromegaly has no ability whatsoever to (directly) control their growth hormone level, not so for the person with a mental disorder. For example, if

Chandra Sripada, *Mental Disorders Involve Limits on Control, Not Extreme Preferences* In: *Agency in Mental Disorder: Philosophical Dimensions*. Edited by: Matt King & Joshua May, Oxford University Press. © Chandra Sripada 2022. DOI: 10.1093/oso/9780198868811.003.0009 one were to put a gun to the head to the person with OCD, they would straightaway desist from washing their hands.

Observations such as these have fueled an alternative perspective that sees mental illness not as a disease, but as a matter of purpose and choice. This "volitional" view has a long history. It is visible in Foucault, Laing, and Szasz (Szasz 1997; Foucault 1988; Laing 1960). It is also seen in newer critiques by Gene Heyman, Hannah Pickard, and Carl Hart (Heyman 2010; Pickard 2012; Hart 2014). The economist Bryan Caplan offers a particularly clear articulation of this volitional position.[1] Using key ideas from consumer theory, Caplan distinguishes constraints *on* actions from preferences *for* actions. He argues physical illnesses produce constraints on one's actions. Mental illnesses do not; they are best understood in terms of volition, albeit in the context of extreme preferences that are out of step with societal norms.

My aim in this chapter is to offer a systematic response to the volitional view of mental illness. The core of my argument is that theorists who support the volitional view operate with a too simple model of human motivational architecture. They view the human mind as having a decision theoretic structure: We have various desires, they differ in strength (reflecting strength of preference), and we always do what we most prefer. I argue the human mind instead has a regulatory control structure. We not only have desires (or similar spontaneous states; I use the term "desire" to refer to all these states for the time being), we have regulatory mechanisms that enable us to modulate or suppress our desires. The presence of regulatory mechanisms introduces the possibility of constraints: if regulation is limited in some way, then certain "lesser" desires that do not reflect what we most prefer may still manifest in action. This is in fact what happens, I argue, in many mental disorders—these disorders arise precisely where the limits of control are breached (in interestingly different ways in different disorders). If this picture is right, then a person's disorder-associated behaviors might not reflect what they most prefer to do, but rather what they are constrained to do.

This chapter is divided into three parts. Section 8.2 distinguishes two models of motivational architecture, the Decision Theory model and the Regulatory Control model. Section 8.3 adopts the Regulatory Control model and sketches a general picture of several major mental illnesses. They are,

---

[1] Caplan 2006. My interest in Caplan's article was spurred by Scott Alexander's discussion on the Slate Star Codex Blog (https://slatestarcodex.com/2020/01/15/contra-contra-contra-caplan-on-psych/).

I argue, conditions that arise due to limits on control. Section 8.3 returns to the key distinction between preferences and constraints. It is argued that an explanation of mental illness based on limits on control is a better overall fit to the data than the volitional view.

8.2 Two Models of Motivational Architecture

8.2.1 The Decision Theory Model

There is a picture of motivational architecture that is extremely common in philosophy, economics, and certain social sciences. The picture resembles a psychologized version of rational choice theory, and it goes like this: People have various desires directed at different things. These desires differ in terms of *strength* (Mele 1998). That is, there are certain motivational properties of these desires in virtue of which they are ordered in terms of motivational "potency" (barring ties—I ignore this complication going forward). Action selection systems are configured so that they are sensitive to the strength properties of one's overall set of desires, and the desire that sits atop the strength ordering becomes the basis for action. The explanations for action supplied by this model are simple and intuitive, for example: Joe's desire to go to the movies is stronger than his desire to do anything else (go to the park or go to the mall, etc.), and so Joe goes to the movies.

This picture of motivational architecture, which I will call the Decision Theory view is so widespread in philosophy and economics, it hardly gets noticed or mentioned. It simply serves as the background default view for understanding agents. But the view implies two principles that are worth pausing to highlight.

First, because of the way the Decision Theory architecture links one's strongest desire to action, the architecture implies that what an agent does will conform to the following law-like generalization, which has been dubbed the Law of Desire:

Whenever a person acts intentionally, they do what they are most strongly motivated to do at the time.[2]

---

[2] See Mele 2003; Sripada 2014; Barnes 2019 for discussion. The principle as stated is susceptible to counterexamples, but these counterexamples are not relevant for our present purposes. Thus, I prefer this simpler formulation for the present purposes.

The second principle, which is a direct consequence of the first, is what we can call the Law of Revealed Desire:

> Whenever a person acts intentionally, what they do reveals what is their strongest motive.

That is, when a Decision Theory agent acts, we can "read off" from their behavior what they most want. Extending this second principle to mental illness, the volitional view naturally follows. Suppose someone touches a doorknob and washes their hands a dozen times in a row, and now they are washing their hands (intentionally) for the thirteenth time, causing serious skin fissures and substantial pain. The second principle implies that washing their hands this thirteenth time is what they most wanted. To be sure, they have unusual wants, or what Caplan calls "extreme preferences." But, if we assume a Decision Theory architecture correctly describes human motivation, then since this is what they intentionally do, we can be confident this is what they most wanted to do.

8.2.2 Regulatory Control Model

I now turn to an alternative picture of motivational architecture. As we go about ordinary life, various kinds of *spontaneous tendencies* arise. Our attention is grabbed by features of the environment. Habitual action tendencies are elicited. Memory items are spontaneously called to mind. We are "pulled" to think about certain topics. A hallmark of spontaneous tendencies such as these is that they operate as a default—the spontaneous tendencies will manifest in action unless something intervenes to block them.[3]

Such intervention is possible because humans have unique abilities for top-down regulation. In what follows, I discuss these regulatory abilities in two steps. First, I discuss regulation of simple, brief spontaneous tendencies of the kind just considered. Second, I discuss regulation of more complex, temporally extended states such as emotions and cravings.

The regulation of simple, brief spontaneous tendencies is called *cognitive control*, and it is extensively studied in cognitive and clinical neuroscience. A standard method involves study of "conflict tasks." The hallmark of these

---

[3] In Sripada (2020), I discuss the nature of these simple spontaneous tendencies in some detail.

tasks is that they set up a conflict between the simple spontaneous tendencies previously discussed and a second type of motivational state, one's *goals*. These are relatively stable motivational states that are closely connected to one's conscious reflective judgments. Here are three examples of conflict tasks:

Stroop Task (Stroop 1935) – On each trial, subjects are shown a color word ("red," "blue") which is itself printed in an ink color. Subjects are asked to state the ink color of the word on all trials. On congruent trials, the word's meaning and ink color match and it is relatively easy to get the right answer. On incongruent trials, the word's meaning and ink color are discrepant, and subjects must exert control over their spontaneous tendency to read the word, in order to select the correct response.

Go/No Go Task (Donders 1969) – On each trial, subjects see a letter on the screen. Subjects are asked to press a button only if the letter is not "X" and withhold the button press if it an "X." Most of the letters are not "X," for example 90% not "X" to 10% "X." This skewed ratio leads to the development of a habit for button pressing. On trials where the stimulus is not "X," the button pressing habit facilitates correct responding. On "X" trials, subjects must suppress this habit.

Think/No Think Task (Anderson and Green 2001) – During a practice session, subjects are trained to recall pairs of words (e.g., ROACH – ORDEAL; GUM – TRAIN). In the test session, they are given the first member of the pair. They are told that if the word appears in green ink, they are to think about the paired word. If the word appears in red ink, they must not think about the paired word. This requires that they suppress the spontaneous tendency to recall the associated word.

These tasks illustrate that based on their goals, people can perform *control actions*—rapidly executed intra-psychic actions that inhibit, suppress, or otherwise modulate various kinds of simple, brief spontaneous tendencies. Different kinds of control actions target different psychological systems. As a result, people can control a diverse array of simple, spontaneous tendencies, including those associated with attention, memory, thought, belief formation, evaluation, and action selection.

Turn now to complex, "hot," temporally extended spontaneous states such as emotions and cravings. We can regulate these states as well in accordance with our goals. Theorists call this capacity various names including "effortful control," "volitional regulation," and "emotion regulation"

(Gross 1998; Rothbart et al. 2003; Sripada et al. 2014). This kind of regulation is illustrated vividly in fMRI studies of craving regulation (Brody et al. 2007; Kober et al. 2010; Hare, Camerer, and Rangel 2009). In these studies, subjects, for example smokers or dieters, are shown pictures of stimuli (cigarettes, indulgent food, etc.) that are known to elicit strong cravings. On some trials, they are asked to simply experience the cravings. On other trials, they are asked to regulate the cravings and reduce their intensity. This is usually accomplished by attention control actions (directing attention away from pictures) and thought control actions (intentionally inhibiting certain thoughts or bringing to mind competing thoughts). These studies typically find:

1) elevated activation in reward-related regions during experience trials;
2) elevated activation in "executive" regions during regulation trials; and
3) an inverse relationship between activity in executive regions and reward regions (suggesting the former is inhibiting the latter).

It is an interesting question how regulation of complex spontaneous states such as emotions and cravings relates to cognitive control, i.e., regulation of the simple, brief spontaneous tendencies I discussed earlier. I discuss this issue in detail elsewhere (Sripada 2020). In short, I think the two are related as whole and part: When a person regulates complex states, they perform a sequence of cognitive control actions directed at simple, brief spontaneous tendencies. I put this issue aside for our present purposes.

I need a general term to refer to this broad collection of spontaneous states, either simple or complex, irrespective of whether they pertain to belief, memory, thought, or action selection. I refer to them all as "pulses." I also need a term to describe different forms of goal-directed regulation, spanning cognitive control over simple, brief states and more complex forms of regulation over complex states. Going forward, I refer to them all as "regulatory control," or "regulation" for short.

Recall the two principles that characterize the Decision Theory agent, the Law of Desire and the Law of Revealed Desire. Critically, they need not hold in a Regulatory Control agent if one additional condition is met: regulation is limited. If regulation is in some way inefficient, weak, or fallible, then an agent can do things that they themselves do not most want to do,[4]

---

[4] Importantly, what an agent "most wants" is determined by motivational properties of desires, not by observing which desire actually manifests in action. See Mele 2003; Sripada 2014 for further discussion.

in violation of the Law of Desire. This will happen when three conditions hold:

1) A person's strongest overall desire is to do one thing (e.g., pay attention to a lecture),
2) They experience spontaneous pulses to do something else (e.g., to notice the ticking of the clock or to mind wander onto some meaningless topic).
3) Top-down regulation is in some way limited, allowing spontaneous pulses to manifest in action (recall that pulses are motivational defaults and thus will be the basis for action unless they are regulated).

When these conditions hold, the person will act on pulses rather than on what they most want, in violation of the Law of Desire. It follows that what they do also fails to reflect what they most want, in violation of the Law of Revealed Desire.

Now, the details here are complex because with a Regulatory Control agent, there are different, somewhat independent sources of motivation arising from the states that I have been calling goals and pulses. Thus, the notion of "what an agent most wants" is more challenging to define: Is it one's strongest goal? One's strongest pulse? Can their respective strengths even be compared? I do not want to get bogged down in these details.[5] It suffices for our purposes to take note of the fact that with a Regulatory Control agent, even if they have the sincere goal of doing one thing, due to limitations on regulatory control, they can still end up doing something else.

8.2.3 Humans Have a Regulatory Control Architecture

There is extensive evidence, reviewed elsewhere (Botvinick and Cohen 2014; Cohen 2017; Hofmann, Schmeichel, and Baddeley 2012), that human motivational architecture has a regulatory control structure, and that is what I will be assuming going forward. Notice, though, that even if the Regulatory Control model is correct, the Decision Theory model remains useful. For example, in most ordinary contexts where there is no need for regulation, or where regulation is so easy it operates flawlessly, then the Decision Theory

---

[5] I discuss this issue in some detail in Sripada (2014).

model and Regulatory Control model will yield similar behavioral predictions. So, the Decision Theory model represents a simplification that works fairly well for day-to-day purposes. However, and this is critical, the two models do come apart in some contexts, and mental illness, as I will presently argue, is a striking example.

8.3 Mental Illness and Dyscontrol

Having introduced the Regulatory Control model of motivational architecture, I now want to fill in the details of how, given this architecture, regulatory control fails in ways relevant to mental disorders. I refer to a state in which the limits of control are breached as a *dyscontrol state*. At a highly general level, all dyscontrol states arise from a mismatch between regulation efficacy and pulse efficacy, which in turn arises from one of three possibilities: an elevated "load" of pulses, a decrease in the person's regulatory capacities, or both. Once we move past this generalization, however, and look at specific mental disorders, we find a variety of types of mismatches that are operative. These mismatches involve different types of pulse states (e.g., attentional, emotional, doxastic), different types of decreases or impairments in regulatory capacities, and different types of environmental contexts in which the pulse/regulation mismatches unfold. I will discuss four disorders to illustrate this variety. Along the way, I highlight certain Control Limiting Factors that arise in these disorders that illuminate specific and interestingly different pathways by which the limits of control are breached. An important theme that emerges in what follows is that the Control Limiting Factors that are relevant to psychiatric disorders involve "extended limits"—limits observable only across days, months, and even years.

8.3.1 Obsessive Compulsive Disorder (OCD)

Individuals with OCD have obsessive thoughts, which are typically directed at characteristic themes (e.g., contamination), and these thoughts arouse substantial anxiety and tension. They additionally have repeated urges to perform behaviors related to these obsessive thoughts, for example urges to wash their hands. Importantly, these thoughts and urges do not just happen occasionally, for example a few times a week or several times a day. Rather, they typically occur with much greater frequency: dozens to hundreds of times a day, often occupying a significant portion of the day.

Consider an individual OCD thought (e.g., the thought that one's hand is contaminated) or an individual OCD urge (e.g., the urge to wash one's hand). Each one of these is readily susceptible to regulation. The person can use the regulatory repertoire discussed in the previous section to redirect attention, suppress problematic thoughts, inhibit inappropriate action tendencies, and so on. Regulation will tend to fail, however, if a person experiences densely recurrent thoughts and urges throughout the day, day after day, month after month. Under these circumstances, regulation starts to become too burdensome for the person.[6]

One kind of burden is experiential. The exercise of top-down regulatory capacities is associated with a distinctive effortful phenomenology that is aversive or otherwise negatively valenced (Shenhav et al. 2017). Thus, spending significant stretches of one's day engaging in top-down regulation of thoughts and urges burdens the person with prolonged dysphoric feelings.

A second kind of burden arises from opportunity cost. Top-down regulation is a member of a larger set of cognitive functions called executive functions (Diamond 2013). Other members include planning, deliberation, and high-level problem-solving. Executive functions are underpinned by a shared, or importantly overlapping, set of brain mechanisms that exhibit limited capacity—engaging these executive mechanisms for one purpose entails, for the most part, giving up their use for other purposes.[7] It follows that if a person must engage in top-down regulation for significant stretches of their day, they must pay substantial opportunity costs in foregoing a range of other valuable executive activities—planning, deliberation, problem-solving—that they could have otherwise undertaken.

In short then, the Control Limiting Factor that operates in OCD involves *cumulative burden*. No thought or urge in the disorder is, by itself, particularly hard to control. But when we consider them in their temporal totality—that is, when we consider the cumulative burden of having to regulate all of these densely recurrent thoughts and urges over extended stretches of time, the burden on the person is excessive and regulation predictably falters.

---

[6] I discuss burdens of regulation in OCD in Sripada (forthcoming).

[7] There is substantial evidence, especially from neuroimaging, of a single domain-general executive network (Duncan and Owen 2000; Niendam et al. 2012; Cole and Schneider 2007). The limited capacity of this network is supported by a number of lines of evidence, see Baddeley 1996; Kurzban et al. 2013 for partial reviews.

The obsessional thoughts and urges in OCD illustrate a more general phenomenon that I claim is found in most mental disorders. We see in OCD three key features: (1) a massive population of pulse-type states; (2) the pulses are in some recognizable sense abnormal; (3) the presence of these abnormal pulses is a long-term feature of the person's psychology. Going forward, I refer to this cluster with a convenient short-hand name: "CAPPs," for *chronic aberrant populations of pulses*. Giving the phenomenon a name will, it is hoped, make it easier to recognize just how ubiquitous it is across a wide range of psychiatric disorders.

8.3.2 Attention-Deficity/Hyperactivity Disorder (ADHD)

In ADHD, we see CAPPs, but rather than obsessive thoughts and impulses, the CAPPs pertain to attention.[8] As we transact with the environment, features of the environment "call out" for our attention (Corbetta and Shulman 2002; Serences et al. 2005): a whisper in the hallway, the text message that may have shown up on one's phone, one's own internal spontaneous musings and mind wanderings. This is true for all individuals, with or without ADHD—attentional pulses impinge on the psyche day in and day out.

Most of these attentional pulses do not present much of a problem because we can regulate them, thus staying on task and avoiding inappropriate distraction. Moreover, unlike OCD, regulating attentional distractors is not particularly effortful or dysphoric, and thus it does not create a cumulative burden on the person. In ADHD, however, a problem arises because there is a regulation/pulse mismatch: either attentional pulses are too frequent or regulation efficacy is diminished,[9] leading to a higher than typical failure rate in which inappropriate attentional pulses more frequently "get through." To be clear, individuals with ADHD *can* regulate attentional pulses, and indeed they succeed most of the time. The problem they face is instead statistical. The modern world is unforgiving in placing demands on

[8] I am focusing here on ADHD, inattentive type, the most common type in adults with ADHD. A broadly similar account could be given of ADHD hyperactive type and ADHD impulsive type, where the role of attentional pulses is replaced by motoric pulses and reward-seeking/appetitive pulses, respectively.

[9] There are few attempts to distinguish which of these two factors predominates in ADHD (cf., Friedman-Hill et al. 2010). However, at least some individuals with ADHD have more wide-ranging difficulties with executive functions suggesting that for them, the top-down factor is more heavily implicated.

our attention; tasks and projects at school and at work require unerring focus to get done well, or get done at all. A higher error rate in regulating attentional distractors is enough to create mistakes, forgetfulness, and disorganization—the core symptoms of ADHD. The main Control Limiting Factor that is operative in ADHD is *unreliable control*. The point probability of successfully regulating each attentional pulse remains quite high. But because attentional pulses are so ubiquitous, the person still experiences regular errors, which in turn produce serious negative academic, occupational, and interpersonal consequences.[10]

8.3.3 Major Depressive Disorder

The hallmark of major depression is the presence of the emotion sadness— not mild sadness that is temporary, but severe sadness that is persistent (i.e., on most occasions for an extended duration). Emotions such as sadness produce a multitude of effects on one's psychology that are mediated by pulse-type states, as I have argued elsewhere in detail (Sripada 2020). Here are some of sadness's effects (Freed and Mann 2007; Hybels et al. 2009; Cipriani et al. forthcoming; Gaddy and Ingram 2014): One's attentional patterns are changed—negative or potentially threatening features of the environment now spontaneously draw one's attention. One's spontaneous interpretations change—ambiguous or neutral events are now interpreted in a negative or pessimistic light. One's thoughts change—negative memories about the past or pessimistic prospections about the future spontaneously enter one's mind. One's action tendencies change—there is a pervasive sense of fatigue that makes doing even basic things feel overwhelmingly effortful. In short, then, depression is a condition that involves chronic alterations in pulses arising from multiple psychological systems; that is, it involves CAPPs.

As in the other cases, these pulses associated with attention, belief, thought, or action can be regulated. For example, a person can suppress a negative memory or force themselves to get out of bed on any particular occasion if supplied with sufficient incentives. The relevant question, however, is whether in real world circumstances where such salient incentives are absent and these problematic pulses arise nearly continuously, can an ordinary person without specialized training in higher-order control

---

[10] I discuss fallibility in the context of cumulative risk of relapse in addiction at length elsewhere (Sripada 2018).

regulate them? The answer is "no" and for multiple reasons. One factor is *deference*. Ordinary people's default position is to accept their spontaneously formed beliefs and impressions. It is rare for people to take a meta-cognitive stance and check carefully whether the way things seem corresponds to the way things actually are. A second factor is *vigilance failure*. Without specialized training in sustained meta-cognitive monitoring, an ordinary person cannot stand at guard monitoring and regulating their own ongoing beliefs, impressions, and thoughts continuously. A third factor is *lack of regulatory skill*. Suppose a person does manage to recognize that something is "off" about an impression that arises on a particular occasion—say, the impression that nobody likes them. Simple suppression strategies might succeed in pushing the thought out of their mind for a moment, but thoughts such as these often immediately return.

Now, there are more sophisticated ways to defeat such thoughts. For example, cognitive behavioral therapy (Aaron T. Beck 1963, 1964, 1979) trains a person to systematically challenge the evidential basis of problematic automatic thoughts, so that undermining these thoughts becomes routinized and more permanent. Advanced meditative training seeks to impart comprehensive control over how attention is directed and how thoughts arise (Rubia 2009). Skills such as these, however, are an *achievement*; they are attained by relatively few, and they are not something that ordinary people simply execute as a matter of course. *Deference*, *vigilance failure*, and *lack of regulatory skill* might each be considered Control Limiting Factors take alone. When they operate together, they surely constitute limits on one's control.

8.3.4 Schizophrenia

In schizophrenia, we once again see the operation of CAPPs. According to a leading theory, the central cognitive/motivational alteration in schizophrenia is abnormal salience.[11] *Salience* refers to a property of a stimulus to grab attention and become the target of valenced appraisal. In schizophrenia, ordinary day-to-day stimuli acquire inappropriate hypertrophied salience: a smile by a stranger, two people coincidentally sharing the same name, a dog with a distinctive limp. These events are passed over in neurotypical individuals, but in individuals with schizophrenia, they strike the person as deeply

---

[11] Kapur 2003; Howes and Kapur 2009. The theory actually pertains to psychosis. Schizophrenia is more complex syndrome with psychosis as a central element.

important and self-relevant, and they become the targets of spontaneous interpretative activity to try to make sense of them. Over time (typically years), ongoing interpretive activity targeting countless events and situations crystallizes in the formation of a delusional system, a system of internally coherent beliefs that makes sense of the person's subjective experience.

Now, for most people, the formation of odd, bizarre beliefs—ones that that are wildly out of step with one's other beliefs about the world and that are not shared with others in one's cultural milieu—are noticed by the person (De Neys and Glumicic 2008; Mercier 2020). This in turn generates efforts, mediated by executive systems (i.e., systems that implement top-down regulatory control), to challenge and correct the errant beliefs. Strikingly, this does not happen in schizophrenia. Thus, a second factor is likely at work: reduced monitoring. Ongoing surveillance of beliefs, already somewhat lax in neurotypical individuals, is compromised still further in schizophrenia.[12] Thus, errant beliefs evade executive correction processes and remain in place, and over time, they become entrenched.

In short then, schizophrenia is a disorder whose etiology is rooted in CAPPs. Abnormalities in salience lead to ongoing bombardment with "doxastic pulses": spontaneous appraisals of day-to-day events in distorted (often paranoid) ways. Many of the Control Limiting Factors discussed earlier likely play a role in explaining why these pulses are not regulated: ordinary people are excessively deferent to their spontaneous impressions; inappropriate doxastic pulses overload executive correction mechanisms; people are inexpert at challenging ill-founded beliefs. And there are likely additional factors, such as impaired monitoring of errant beliefs, that are operative in schizophrenia specifically.

8.3.5 Summing Up

In this section, I discussed four major psychiatric disorders. My analysis of what goes on in these disorders had a common structure: All these disorders centrally involve chronic aberrant populations of pulses, or CAPPs, and, in some cases, there were also inefficiencies or impairments in regulatory

---

[12] Dopamine dysfunction provides a unifying explanation of why the two deficits are paired: midbrain dopamine pathways are involved in salience processing (Kapur 2003) while mesocortical dopamine pathways are involved in executive functions, which include monitoring (Braver, Barch, and Cohen 1999; Goldman-Rakic et al. 2004).

capacities. In each disorder, the disorder-associated pulses *taken in totality, typically over long periods of time* breach certain limits of control, thus explaining why the person exhibits the characteristic disorder-associated symptoms. Space does not allow me to discuss more disorders or conditions, such as addiction, mania, or anxiety disorders. But the general form of how I would explain these conditions is already clear. In short then, on my view, a key feature of many major mental disorders is that they involve limits on control.[13]

8.4 Preferences or Constraints Revisited

Are mental disorders best explained in terms of limits on control, or do they reflect volition in the setting of extreme (and socially stigmatized) preferences? I now want to do some argument "scorekeeping" comparing the two views, focusing on some of Caplan's arguments.

8.4.1 Incentive Sensitivity and the "Gun to the Head Test"

One of the main arguments for the volitional view involves the "gun-to-the-head-test." Caplan, for example, writes:

> Can we change a person's behavior purely by changing his incentives? If we can, it follows that the person was able to act differently all along, but preferred not to; his condition is a matter of preference, not constraint . . .
>
> (Caplan 2006, 349)

Here Caplan presents crisply and succinctly what is probably the most common theme in a vast "anti-psychiatry" literature: mental disorders involve choices that are stigmatized, but there is no genuine loss of control or impairments in agency. Variants of the gun-to-the-head test (or more general incentive sensitivity tests) are put forward by Pickard, Hart, Heyman, Morse, Foddy and Savalescu, and many others (Pickard 2012; Hart 2014; Heyman 2010; Morse 2002; Foddy and Savulescu 2010).

---

[13] This weaker claim is what I need for the present purposes in critiquing Caplan. I actually endorse a stronger view: There is a deep conceptual tie between mental disorder and dyscontrol, and thus *all* mental disorders involve limits on control. I will develop this view in due course, but I do not try to defend it here.

We are now in a position to see why conclusions based on the gun-to-the-head-test are misleading. In my account of mental disorders, I emphasized the role of chronic pulses, i.e., CAPPs, and I identified a number of Control Limiting Factors that arise specifically in that context, which in turn lead to the characteristic thoughts and actions we see in these disorders. The gun-to-the-head test, however, describes a scenario with little to no relevance to mental disorders because CAPPs are absent and none of the Control Limiting Factors have a chance to operate.

Consider a person with OCD. They can stop washing their hands if you put a gun to their head. But they still face limits on control that arise from the *cumulative burden* of having to regulate an unending, recurrent series of dysphoric urges. If you put a gun to the head of someone with ADHD, they can regulate a certain distracting attentional pulse (indeed they succeed at this anyways most of the time). Their problem is one of *unreliable control* in the context of temporally extended projects, and the gun-to-the-head-test has nothing to say about this. A person with depression can interpret a situation less negatively if you threaten them with certain death. But this threat is explicit and, by stipulation, definitive. In their day-to-day lives, however, they need to regulate ongoing negative interpretations and thoughts that lack this kind of clarity and certitude, allowing *deference, overload*, and *lack of regulatory skill*, among other Control Limiting Factors, to operate. A person with schizophrenia can be ordered under threat of serious harm to re-interpret events in less paranoid ways. But such interventions, if they work at all, are invariably temporary. Due to continued aberrant salience attribution and *impaired monitoring* of errant beliefs, among other Control Limiting Factors, spontaneous paranoid interpretations will soon return and their delusional system will be reinstated.

The gun-to-the-head test initially strikes us as plausible because we have a picture that when agency breaks down, barriers to purposive actions are decisive, rigid, and easy to see with a quick look. For example, in contrasting mental illness and physical illness, Caplan notes that no incentive can get someone who is paralyzed to stand (Caplan 2006, 342). Here the absence of incentive sensitivity is clear with a single glance. Many theorists similarly seem to assume that mental disorders need to impair agency in a similarly decisive and easy to check way. But dyscontrol in mental disorders is, as we have seen, not much like this at all. It instead involves temporally extended faltering of agency due to the cumulative impact of CAPPs, with substantial preserved incentive sensitivity at any given slice of time.

Now, to be absolutely clear, I am not arguing that, in contrast to physical disorders, mental disorders yield only weak constraints on thought and action. That is actually the opposite of my view; I believe mental disorders produce constraints on thought and action that are serious and severe. A person bombarded with obsessive thoughts of contamination and urges to hand wash is in a very real sense coerced (intra-psychically) into doing what they do. My point is that with mental illness, there is not one single or even several decisive blows that can be easily spotted, but rather countless tiny cuts that may be much harder to appreciate.

8.4.2 A Unified Model of Human Behavior

Another claimed advantage of the volitional view is that it presents a unified model of behavior. All purposive behavior, both healthy and disordered, is explained as arising from one's ordinary preference-based motivational psychology. Caplan complains that economists have been too willing to carve out a special exception for mental illness, as if the laws of preference-based behavior apply everywhere else but somehow not there. He writes:

> Though these authors are usually eager to bring social phenomena into the orbit of economics, they not only make an exception for severe mental illness; they treat the exception as uncontroversial. Over time, however, diagnoses of mental illness have become increasingly widespread. Epidemiologists now report that 20% or more of the USA population suffers from mental illness during a given year (Kessler et al. 1994). A seemingly small loophole in the applicability of economics has grown beyond recognition. (Caplan 2006, 334)

The view that mental illness involves limits on control, however, avoids Caplan's charge because the view invokes a single model of motivation for all behavior: the Regulatory Control model. Most ordinary behavior arises within the "regulation frontier" of the architecture: regulation works properly either because it is not needed (the relevant pulse states are situationally appropriate) or because it succeeds in subduing problematic pulse states. In some cases, the limits of control of this Regulatory Control architecture are systematically breached and the person regularly exhibits dyscontrol characteristic of a psychiatric illness. But there are not two models of behavior here. There is a single model that involves multiple parameters (e.g., efficacy of pulses, efficacy of regulation, etc.), and health and disease occupy different regions of the parameter space. Just like a model of a car engine explains both why a Mustang hums and why it sputters, the Regulatory Control model explains purposive agency in both health and disease.

8.4.3 Dystonicity

I now turn to a feature of mental illness that is hard for the volitional view to explain but makes perfect sense with the limits on control model.

Many mental disorders are "ego dystonic": The person repudiates, rejects, or in some other way "stands against" their disorder-associated thoughts and actions (Freud 2014; Clark 1992; Belloch, Roncero, and Perpiñá 2012; Purdon et al. 2007). To be sure, some disorders are not dystonic in this way, at least overtly. For example, people with paranoid schizophrenia do not typically come to the doctor seeking out help with their delusions. But with many disorders, e.g., OCD, ADHD, and depression, people with the conditions actively seek out clinical care and pursue fairly demanding treatments. A natural explanation for why they do this is that there is something about their thoughts and actions that they dislike and want to change. But this natural interpretation makes little sense on the volitional view of mental illness.

To make this point concrete, take a person with ADHD. According to supporters of the volitional view such as Caplan, this person most prefers to chase variety and distraction—that is why they are disorganized, forgetful, and scattered. If that is truly their strongest preference—that is, if their preference ranking really is *chasing variety/distraction > being organized*—then it is puzzling why they are at the clinic month after month working with a psychiatrist on a medication regimen and working with a behavioral therapist on extensive cognitive/behavioral treatments.

Defenders of the volitional view might respond that we need to distinguish what the person herself prefers from societal reactions and stigma. While the person herself most prefers variety and distraction, she is nonetheless at the clinic to change her thoughts and actions because that is "what society demands." But this response falters because it relies on an inappropriately restrictive understanding of preferences. If chasing variety and distraction is tightly linked to the emergence of interpersonal problems for the person, then we need to change the descriptions of their options to reflect this. We thus assess their preferences over the following "conjoined" outcomes: *chasing variety/distraction and incurring interpersonal problems* vs. *not chasing variety/distraction and not incurring interpersonal problems*. If the person prefers the latter, then they do not have a problem according to the Decision Theory model of motivation, i.e., the model that undergirds the volitional view. They will just straightaway not chase variety and distraction and avoid the interpersonal problems that would have ensued. But if they prefer the former, then we are back to our original problem: Why are they in the clinic week after week undertaking costly and burdensome treatments to rid themselves of thoughts and actions that, according to the volitional view, they actually genuinely prefer? The volitional view does not seem to have a good answer.

The view that mental illness involves limits on control view, on the other hand, has a ready explanation for dystonicity. The person with ADHD is in the clinic week after week because she has the goal of being organized and timely and thereby achieving all the positive consequences that flow from that (occupational and interpersonal success, etc.). But she is beset by distracting attentional pulses that arise irrespective of these goals, and, though she can regulate many of these attentional pulses, she cannot successfully regulate all of them—that is, she has reached the limits of control. So, she now finds herself doing all sorts of things—for example, being forgetful and disorganized—that she does not really want to do. The basic form of this explanation generalizes to a wide range of psychiatric disorders. In addition to OCD and depression (discussed earlier), it extends to other conditions, such as anxiety and addiction, where, though I did not discuss them, it is not hard to see how to apply the general form of this model.

Stepping back a bit, the fundamental problem for the volitional view of mental illness is that to explain dystonicity that clearly attends many mental disorders, we need a way for agents to regularly and recurrently do things that they prefer not to do, even hate to do (e.g., wash their hands for the 100th time or have lapses of attention for the thousandth time). The volitional view, however, relies on the Decision Theory model of motivation. As such, it obeys the Law of Desire, which says roughly that agents do what they most want to do. But by tying action so tightly to preference, this law seems to make dystonicity, especially chronic dystonicity of the kind seen in psychiatry, impossible.[14]

---

[14] One move available to supporters of the volitional view is to appeal to "meta-preferences" (see for example Caplan's blog post "The Depression Preference" (https://www.econlib.org/the-depression-preference/). The idea is that a depressed person prefers to lie in bed and think guilty

8.5 Conclusion

Consumer theory distinguishes between one's preferences, what one wants to do, and one's budget, what one is able to do. The volitional view of mental illness locates mental illness on the preference side—mental illness involves choice rather than constraints on what one is able to do. The choices are, to be sure, sharply out of step with societal norms and are thus stigmatized, but they remain just that: choices.

In responding to volitional view of mental illness, I put forward a more structured model of motivational architecture, one that countenances both spontaneous states as well as regulatory capacities that are responsive to our goals and that regulate these spontaneous states. But regulation has its limits, especially when it must be deployed over extended intervals of time (months and years) against massive populations of spontaneous tendencies to think and do various things. The existence of limits on control opens up space for agents to regularly and recurrently think things and do things that they themselves prefer not to think and do, and mental disorders, I argued, reside in this space. The constraints on thought and action found in mental disorders are certainly different in kind from constraints in physical conditions. They are, nonetheless, no less real.

References

Anderson, Michael C., and Collin Green. 2001. "Suppressing Unwanted Memories by Executive Control." *Nature* 410 (6826): 366.

Baddeley, Alan. 1996. "Exploring the Central Executive." *The Quarterly Journal of Experimental Psychology Section A* 49 (1): 5–28.

---

thoughts. At the same time, however, they "meta-prefer" to not have depressive first-order preferences and that is why they find their depressive behaviors ego dystonic and seek treatment. However, if we try to combine the meta-preference view with the Decision Theory model of motivation and its associated Law of Desire, the result is incoherence. The problem can be stated in the form of a dilemma. If the first-order preference to lay in bed is the person's strongest, then why is the person at the clinic week after week seeking to defeat this desire? On the other hand, if the meta-preference is the person's strongest, then the basic premise of the volitional view is falsified. The person with depression does *not* most prefer to lie in bed and think guilty thoughts as originally claimed; they actually most prefer essentially the *opposite*. That is, they strongly disprefer having the motives on the basis of which they do these things, and they want those first-order preferences to be eradicated. If we go further and try to explain why the person cannot seem to bring about what they most strongly meta-prefer, the most plausible answer is that they there is some constraint that prevents them. In this way, the meta-preference view quickly ends up abandoning volition in favor of constraints.

Barnes, Eric Christian. 2019. "An Argument for the Law of Desire." *Theoria* 85 (4): 289–311.

Beck, Aaron T. 1963. "Thinking and Depression. I. Idiosyncratic Content and Cognitive Distortions." *Arch. Gen. Psychiatry* 9: 324–33.

Beck, Aaron T. 1964. "Thinking and Depression. II. Theory and Therapy." *Arch. Gen. Psychiatry* 10: 561–71.

Beck, Aaron T. 1979. *Cognitive Therapy of Depression.* New York: Guilford Press.

Belloch, Amparo, María Roncero, and Conxa Perpiñá. 2012. "Ego-Syntonicity and Ego-Dystonicity Associated with Upsetting Intrusive Cognitions." *Journal of Psychopathology and Behavioral Assessment* 34 (1): 94–106.

Botvinick, Matthew M., and Jonathan D. Cohen. 2014. "The Computational and Neural Basis of Cognitive Control: Charted Territory and New Frontiers." *Cognitive Science* 38 (6): 1249–85. https://doi.org/10.1111/cogs.12126.

Braver, Todd S., Deanna M. Barch, and Jonathan D. Cohen. 1999. "Cognition and Control in Schizophrenia: A Computational Model of Dopamine and Prefrontal Function." *Biological Psychiatry* 46 (3): 312–28.

Brody, Arthur L., Mark A. Mandelkern, Richard E. Olmstead, Jennifer Jou, Emmanuelle Tiongson, Valerie Allen, David Scheibal, Edythe D. London, John R. Monterosso, and Stephen T. Tiffany. 2007. "Neural Substrates of Resisting Craving during Cigarette Cue Exposure." *Biological Psychiatry* 62 (6): 642–51.

Caplan, Bryan. 2006. "The Economics of Szasz: Preferences, Constraints and Mental Illness." *Rationality and Society* 18 (3): 333–66.

Cipriani, Andrea, A. Tomlinson, B. Teufer, A. M. Chevance, G. Gartlehner, S. Touboul, P. Ravaud, C. LeBerre, E. I. Fried, and V. T. Tran. Forthcoming. "Identifying Outcomes for Depression that Matter to Patients, Informal Caregivers and Healthcare Professionals: Qualitative Content Analysis of a Large International Online Survey." *Lancet Psychiatry.*

Clark, David A. 1992. "Depressive, Anxious and Intrusive Thoughts in Psychiatric Inpatients and Outpatients." *Behaviour Research and Therapy* 30 (2): 93–102.

Cohen, Jonathan D. 2017. "Cognitive Control." In *The Wiley Handbook of Cognitive Control*, 1–28. Oxford: Wiley-Blackwell. https://doi.org/10.1002/9781118920497.ch1.

Cole, Michael W., and Walter Schneider. 2007. "The Cognitive Control Network: Integrated Cortical Regions with Dissociable Functions." *NeuroImage* 37 (1): 343–60. https://doi.org/10.1016/j.neuroimage.2007.03.071.

Corbetta, Maurizio, and Gordon L. Shulman. 2002. "Control of Goal-Directed and Stimulus-Driven Attention in the Brain." *Nature Reviews. Neuroscience* 3 (3): 201–15. https://doi.org/10.1038/nrn755.

De Neys, Wim, and Tamara Glumicic. 2008. "Conflict Monitoring in Dual Process Theories of Thinking." *Cognition* 106 (3): 1248–99.

Diamond, Adele. 2013. "Executive Functions." *Annual Review of Psychology* 64: 135–68.

Donders, Franciscus Cornelis. 1969. "On the Speed of Mental Processes." *Acta Psychologica* 30: 412–31.

Duncan, John, and Adrian M. Owen. 2000. "Common Regions of the Human Frontal Lobe Recruited by Diverse Cognitive Demands." *Trends in Neurosciences* 23 (10): 475–83. https://doi.org/10.1016/S0166-2236(00)01633-7.

Foddy, Bennett, and Julian Savulescu. 2010. "A Liberal Account of Addiction." *Philosophy, Psychiatry, & Psychology: PPP* 17 (1): 1.

Foucault, Michel. 1988. *Madness and Civilization: A History of Insanity in the Age of Reason.* Vintage.

Freed, Peter J., and J. John Mann. 2007. "Sadness and Loss: Toward a Neurobiopsychosocial Model." *American Journal of Psychiatry* 164 (1): 28–34.

Freud, Sigmund. 2014. *On Narcissism: An Introduction.* Redditch, UK: Read Books Ltd.

Friedman-Hill, Stacia R., Meryl R. Wagman, Saskia E. Gex, Daniel S. Pine, Ellen Leibenluft, and Leslie G. Ungerleider. 2010. "What Does Distractibility in ADHD Reveal about Mechanisms for Top-down Attentional Control?" *Cognition* 115 (1): 93–103.

Gaddy, Melinda A., and Rick E. Ingram. 2014. "A Meta-Analytic Review of Mood-Congruent Implicit Memory in Depressed Mood." *Clinical Psychology Review* 34 (5): 402–16.

Goldman-Rakic, Patricia S., Stacy A. Castner, Torgny H. Svensson, Larry J. Siever, and Graham V. Williams. 2004. "Targeting the Dopamine D 1 Receptor in Schizophrenia: Insights for Cognitive Dysfunction." *Psychopharmacology* 174 (1): 3–16.

Gross, James J. 1998. "The Emerging Field of Emotion Regulation: An Integrative Review." *Review of General Psychology* 2: 271–99.

Hare, T. A., C. F. Camerer, and A. Rangel. 2009. "Self-Control in Decision-Making Involves Modulation of the VmPFC Valuation System." *Science* 324: 646–48. https://doi.org/10.1126/science.1168450.

Hart, Carl. 2014. *High Price: A Neuroscientist's Journey of Self-Discovery That Challenges Everything You Know About Drugs and Society.* Reprint edition. New York, NY: Harper Perennial.

Heyman, Gene M. 2010. *Addiction: A Disorder of Choice.* Reprint edition. Cambridge, Mass.; London: Harvard University Press.

Hofmann, Wilhelm, Brandon J. Schmeichel, and Alan D. Baddeley. 2012. "Executive Functions and Self-Regulation." *Trends in Cognitive Sciences* 16 (3): 174–80.

Howes, Oliver D., and Shitij Kapur. 2009. "The Dopamine Hypothesis of Schizophrenia: Version III—the Final Common Pathway." *Schizophrenia Bulletin* 35 (3): 549–62.

Hybels, Celia F., Dan G. Blazer, Carl F. Pieper, Lawrence R. Landerman, and David C. Steffens. 2009. "Profiles of Depressive Symptoms in Older Adults Diagnosed with Major Depression: Latent Cluster Analysis." *The American Journal of Geriatric Psychiatry* 17 (5): 387–96.

Kapur, Shitij. 2003. "Psychosis as a State of Aberrant Salience: A Framework Linking Biology, Phenomenology, and Pharmacology in Schizophrenia." *American Journal of Psychiatry* 160 (1): 13–23.

Kessler, RC, McGonagle KA, Zhao S, and et al. 1994. "Lifetime and 12-Month Prevalence of DSM-III-R Psychiatric Disorders in the United States: Results from the National Comorbidity Survey." *Archives of General Psychiatry* 51 (1): 8–19. https://doi.org/10.1001/archpsyc.1994.03950010008002.

Kober, Hedy, Peter Mende-Siedlecki, Ethan F. Kross, Jochen Weber, Walter Mischel, Carl L. Hart, and Kevin N. Ochsner. 2010. "Prefrontal–Striatal Pathway Underlies Cognitive Regulation of Craving." *Proceedings of the National Academy of Sciences* 107 (33): 14811–16. https://doi.org/10.1073/pnas.1007779107.

Kurzban, Robert, Angela Duckworth, Joseph W. Kable, and Justus Myers. 2013. "An Opportunity Cost Model of Subjective Effort and Task Performance." *The Behavioral and Brain Sciences* 36 (6): 661–79. https://doi.org/10.1017/S0140525X12003196.

Laing, R. D. 1960. *The Divided Self: An Existentialist Study in Sanity and Madness.* Penguin.

Mele, Alfred. 1998. "Motivational Strength." *Noûs* 32 (1): 23–36. https://doi.org/10.1111/0029-4624.00085.

Mele, Alfred. 2003. *Motivation and Agency.* New York: Oxford University Press.

Mercier, Hugo. 2020. *Not Born Yesterday: The Science of Who We Trust and What We Believe.* Princeton, NJ: Princeton University Press.

Morse, Stephen J. 2002. "Uncontrollable Urges and Irrational People." *Virginia Law Review* 88: 1025–78.

Niendam, Tara A., Angela R. Laird, Kimberly L. Ray, Y. Monica Dean, David C. Glahn, and Cameron S. Carter. 2012. "Meta-Analytic Evidence for a Superordinate Cognitive Control Network Subserving Diverse Executive Functions." *Cognitive, Affective, & Behavioral Neuroscience* 12 (2): 241–68. https://doi.org/10.3758/s13415-011-0083-5.

Pickard, Hanna. 2012. "The Purpose in Chronic Addiction." *AJOB Neuroscience* 3 (2): 40–9. https://doi.org/10.1080/21507740.2012.663058.

Purdon, Christine, Emily Cripps, Matthew Faull, Stephen Joseph, and Karen Rowa. 2007. "Development of a Measure of Egodystonicity." *Journal of Cognitive Psychotherapy* 21 (3): 198–216.

Rothbart, Mary K., Lesa K. Ellis, M. Rosario Rueda, and Michael I. Posner. 2003. "Developing Mechanisms of Temperamental Effortful Control." *Journal of Personality* 71 (6): 1113–44.

Rubia, Katya. 2009. "The Neurobiology of Meditation and Its Clinical Effectiveness in Psychiatric Disorders." *Biological Psychology* 82 (1): 1–11.

Serences, John T., Sarah Shomstein, Andrew B. Leber, Xavier Golay, Howard E. Egeth, and Steven Yantis. 2005. "Coordination of Voluntary and Stimulus-Driven Attentional Control in Human Cortex." *Psychological Science* 16 (2): 114–22.

Shenhav, Amitai, Sebastian Musslick, Falk Lieder, Wouter Kool, Thomas L. Griffiths, Jonathan D. Cohen, and Matthew M. Botvinick. 2017. "Toward a Rational and Mechanistic Account of Mental Effort." *Annual Review of Neuroscience* 40: 99–124.

Sripada, Chandra. 2014. "How Is Willpower Possible? The Puzzle of Synchronic Self-Control and the Divided Mind." *Noûs* 48: 41–74.

Sripada, Chandra. 2018. "Addiction and Fallibility." *The Journal of Philosophy* 115 (11): 569–87.

Sripada, Chandra. Forthcoming. "Loss of Control in Addiction: The Search for an Adequate Theory and the Case for Intellectual Humility." In *Oxford Handbook of Moral Psychology*, edited by John M. Doris and Manuel Vargas. https://umich.box.com/s/soc6hhnz9yexm09r6hg8dugd60opu0um.

Sripada, Chandra. 2020 (online early view). "The Atoms of Self-Control." *Nous*.

Sripada, Chandra, Michael Angstadt, Daniel Kessler, K. Luan Phan, Israel Liberzon, Gary W. Evans, Robert C. Welsh, Pilyoung Kim, and James E. Swain. 2014. "Volitional Regulation of Emotions Produces Distributed Alterations in Connectivity between Visual, Attention Control, and Default

Networks." *NeuroImage* 89 (April): 110–21. https://doi.org/10.1016/j.neuroimage.2013.11.006.

Stroop, J. Ridley. 1935. "Studies of Interference in Serial Verbal Reactions." *Journal of Experimental Psychology* 18 (6): 643.

Szasz, Thomas. 1997. *Insanity: The Idea and Its Consequences.* Syracuse, NY: Syracuse University Press.

Index

Mesh in urogynaecology

❶ This is a contentious and topical issue.

What is mesh?

Mesh refers to manufactured synthetic implantable devices, whereas the term 'grafts' should be used for biological implants. A 'tape' is a thin strip of synthetic material. Implants are used to augment weak connective tissue, for example, in incontinence or pelvic organ prolapse.

Biological grafts

• Autologous grafts—harvested from the patient's own tissue, e.g. rectus sheath, fascia lata
• Allografts—cadaveric tissue
• Xenografts—acellular extracts of collagen, typically bovine or porcine. Use is limited by inconsistency of tissue strength.

Synthetic mesh

• Synthetic absorbable mesh—polyglactin or polyglycolic acid. Animal models show poor long-term tensile strength.
• Synthetic non-absorbable mesh—should provide durable tissue strength.
• Classification is shown in Table 11.1.

Table 11.1 Classification of synthetic non-absorbable mesh

Mesh type	Pore size	Description
Type 1	>75µm	Completely macroporous and monofilamentous
Type 2	<10µm	Totally microporous: pore size smaller than 10µm in at least 1/3 dimensions
Type 3	>75µm	Macroporous with either multifilamentous or microporous components
Type 4	<1µm	Submicronic pores

▶ Type 1 meshes are the preferred option for gynaecological surgery. The larger pore size permits infiltration by fibroblasts, blood vessels, collagen fibres, and macrophages, enabling tissue incorporation and reduced infection risk.

Use of mesh

• Abdominal—sacrocolpopexy
• Laparoscopic—sacrocolpopexy, hysteropexy
• Vaginal—mid-urethral slings, transvaginal mesh for POP.